An advocate of the constructive mathematical viewpoint of Bruno de Finetti, Frank Lad is a research associate at the Department of Mathematics and Statistics, University of Canterbury. He is the author of *Operational Subjective Statistical Methods: a mathematical, philosophical, and historical introduction*, New York: John Wiley, 1996. Many journal publications in mathematical statistics are available on his pages at researchgate.net. These include "Extropy: complementary dual of entropy", *Statistical Science*, 2015 with Giuseppe Sanfilippo and Gianna Agrò. He has lectured widely throughout the world. Residing in Otautahi/Christchurch, Aotearoa/New Zealand, he currently tends his garden, and tutors neighbourhood children in arithmetic.

BA Mathematics, University of Dayton, 1970
MA Economics, University of Michigan, 1972
MA Statistics, University of Michigan, 1973
PhD Econometrics, University of Michigan, 1974

Statement of Purpose:

I have written every detail of this book for mathematicians and for professional participants in theoretical physics. However, it is also meant for appreciation by the generally educated public who are interested in the foundations of science. The exposition presupposes a modicum of familiarity with mathematics. My goal is to alert you to a mathematical error of neglect that has led respected physicists to misleading claims about matters of quantum phenomena. I dispute neither the probabilities of quantum theory, nor the experimental evidence that has been generated. I dispute the mathematical assessment of their implications. The context of the problem addressed dates back to Einstein's challenge to the completeness of quantum theory in the 1930's, which was revitalised by an anomaly proposed by John Bell in the 1960's. This has been discussed and embellished during the subsequent half-century. Although several components of the present book have been published in professional journals of physics, the scientific community is largely in denial, sheltering their vaunted theoreticians from criticism via methods familiar to readers of Thomas Kuhn's *The Structure of Scientific Revolutions*. You are invited to judge for yourself.

Frank Lad

JUST PLAIN WRONG

*The dalliance of quantum theory
with the defiance of Bell's inequality*

AUSTIN MACAULEY PUBLISHERS™

LONDON * CAMBRIDGE * NEW YORK * SHARJAH

A CIP catalogue record for this title is available from the
British Library.

ISBN 9781035830077 (Paperback)
ISBN 9781035830084 (ePub e-book)

www.austinmacauley.com

First Published 2024
Austin Macauley Publishers Ltd
1 Canada Square
Canary Wharf
London
E14 5AA

To Dolores Terwoord Lad and Robert Augustine Lad,
physical chemists of the Manhattan Project,
who fell in love.

They were proud of their brood,
and taught us to think
rather than merely to repeat what we were told.

Contents

Preface

There is a marvellous remark attributed to Wolfgang Pauli in the lore of quantum physics, pertaining to an argument that he disdained. He considered the understanding it proposed was so far off the mark that he mooted "It's not even wrong".

The book you are holding corrects a mistake that does not involve such sophistication. It concerns specific influential arguments regarding the defiance of Bell's inequality that are just plain wrong, embedding a mathematical error that will be understandable to most anyone.

The error has deep consequence, leading the community of quantum physicists to reject the principle of local realism, and to proclaim bizarre features of quantum phenomena that the theory does not actually support. Proponents of the error have now been deservedly feted with the award of the Nobel Prize in Physics, 2022, by a Committee who have been taken in by it. Shamelessly, but with due respect, I make bold to call them all to account.

There is no shame in making a mistake. We all make them regularly, and we live in the dreams we create to cover for them. When we do come to recognise a mistake, either through insight afforded by the absurdity of the constructs we have designed to account for it, or through confrontation by others who have somehow avoided it, the liberation can engender a mixture of elation and embarrassment. This is one of the ways we learn, by making mistakes.

There is shame in denying our mistakes when we are alerted to them, and in actively suppressing their clarification. This is the current situation of contemporary quantum theorists who are wedded to a conception of the theory as the embodiment of startling pronouncements about the mysteriously random structure of matter at the scale of quantum activity. The probabilistic specifications of quantum mechanics are widely regarded as inhering properties that defy standard features of probabilities appropriate to the assessment of mundane matters at the scale of ordinary orders of magnitude: automobile accidents, product warranties, life insurance policies, earthquakes, bank failures, bowling pins, horse races, you name it.

The supposed identification of unusual properties of quantum probabilities stems from a mistaken assessment of the mathematical expectation of a quantity that is imagined to defy what is called Bell's inequality. Nothing mysterious about it, the accepted assessment of this expected value according to probabilities derived from quantum theory is just plain wrong. There is nothing wrong with the probabilities motivated by quantum theorising. It is their misuse in assessing this specific expected value that is mistaken. It is easy to say what is just plain wrong, but it will take this book to explain the details and to discuss related misconceptions. The mistakes are three, and burgeoning.

The first mistake of quantum theorists occurs in assessing the expected value of a quantity that appears to be a linear function of four variables. The simplest exposition of the error occurs in the context of what is known as the CHSH formulation of a quantum experiment involving the polarisation of light. The acronym derives from a famous article by Clauser, Horne, Shimony, and Holt, who devised it. We will identify it specifically in Chapter 1 on the quantum violation of Bell's inequality as a misunderstanding based on a mathematical error of neglect. The error occurs in assessing a *gedankenexperiment*, a thought experiment designed to assess the validity of Einstein's insisted principle of local realism. My exposition of the error was first published in the *Journal of Modern Physics* in 2021.

Technically, the problem arises because account is not taken of the fact that the value of each one of the four variables involved in the inequality statement is constrained in this context to equal a function value determined by the other three. So the quantity needs be identified as a function of only three variables rather than four. Intriguingly, there are four ways in which it can be so represented. Each of the four variables composing Bell's quantity is constrained to equal a function of the other three, in exactly the same form.

In its simplest exhibition, the error of quantum theorists amounts to assessing the expected value of a quantity denoted by the letter "s", as equal to $2\sqrt{2}$. The quantity itself can only equal either -2 or +2. The fact that the erroneous assessment of the expectation is outside of the interval surrounded by these numbers is one form of the touted defiance of Bell's inequality by the probabilities of quantum physics.

The derivation of this result is just plain wrong. Its recognition does not rely on sophisticated forms of mathematics, but it does require a change in perspective from which the physical situation is typically considered. Assessing the expectation correctly addresses the fact that quantum theory explicitly denies the prospect of making the assertions required to identify the expectation *precisely*. Rather, quantum theory supports only a 4-dimensional polytope of probability distributions relevant to "s" which we will be able to visualise. The expected value of s according to distributions within this polytope is surely not $2\sqrt{2}$, but is merely bounded within the interval $(1.1213, 2]$.

This error of neglect compounds itself when affecting not only the assessed value of this expectation, but also when infecting the widely reported assessment of empirical results that purport to provide experimental evidence of the inequality's defiance. This mistake lies at the base of Alain Aspect's original analysis of experimental optics, and has been followed routinely over several years in the analysis of subsequent experimentation. It runs through all continuing research that has culminated in the proclamation of "Death by experiment to the principle of local realism", announced in the esteemed journal *Nature* in 2015. That claim was based on sophisticated experimentation conducted at the Technical University of Delft. The error also runs through the analysis of all related experimental data.

The very same mistake of neglecting functional relations embedded in a gedankenexperiment arises in a hidden context in a popular article of David Mermin, which has fascinated readers for several decades. It was designed specifically to be readable by anyone, even those untutored in issues of quantum physics. He noted in concluding it that the real context of the problem involves the propagation of two electrons past a pair of magnets, each of which can be positioned in three different rotated angles perpendicular to an incoming electron. We will recognise the error and resolve it in Chapter 4.

A second mistake arises as an error of propositional logic. It occurs in a very influential paper of Greenberger, Horne, Shimony and Zeilinger in the early 1990s, the results of which have been absorbed as gospel in quantum theoretic dogma. This proposes to identify the defiance of Bell's inequality in a 4-dimensional problem without relying on inequalities at all. The authors conclude that it is impossible to construct a model of Einstein's principle of local realism in a problem of more than two dimensions without generating a contradiction. We will find in Chapter 2 that the contradiction "discovered" by the authors was introduced into their analysis by themselves! Studied seriously in detail by a minimum of five thousand professional physicists over the course of more than thirty years, it is surprising that this error of simple logic has not been identified in such an influential paper heretofore. Evading their contradiction, we will find that a sensible assessment of the experiment they propose merely identifies an array of symmetries which it involves.

My assessment appearing here in Chapter 2 has been published originally in the journal *Entropy* in 2020. The version of the chapter appearing here makes some emendations to the published article: changing the equation numbering system to match that of the original GHSZ article, changing the style of source-referencing to one that is more informative to the reader, and transferring a distracting side-issue that is discussed in the published text to an Appendix.

A third mistake running through physics literature concerns Einstein's assertions about the incompleteness of quantum theory and his proposal of supplementary variables as an explanation for its probabilistic results. There is a sizeable body of analysis touting the supposed impossibility of formalising such a proposal, stemming from the initial arguments of John von Neumann, which have since been superseded. Nonetheless, the mathematical structure of the "hidden variables" argument is explained simply and clearly here in Chapter 3, exposing its potential applicability to any proposed quantum distribution whatsoever. A large portion of this was published in the journal *Applied Mathematics* in 2022. The commonly accepted notion that some distributions could be amenable to a hidden variables interpretation, and others not, is simply false. Moreover, in the context of explaining the formalisation, it becomes clearly evident once again that Bell's inequality is not defied in any way by quantum probabilities associated with the gedankenexperiments that motivated its formulation.

Chapter 5 presents the results of a Monte Carlo computer simulation. It has been designed to clarify still another mistake made in the empirical assessment of the inequality defiance, on the basis of averaging the results of many experiments. The thought experiment providing context for the inequality relies on simultaneous observation of paired photon (or paired electron) behaviour under four distinct conditions that cannot possibly be instantiated. Thus, quantum theory itself says nothing precise about its result, and experimental physics cannot produce data relevant to it. It can be resolved only by a simulation. Claims to the contrary have been made on the basis of averages of pseudo-observations that are imagined in such a setting, garnered from sequences of real observations made on *different* pairs of photons under single conditions. These have been championed in publications by Richard Gill, following a tradition that runs through notable works by Alain Aspect, Michael Redhead, and N. David Mermin. A creditable evaluation of their argument requires a simulation study which the chapter provides. It will have to speak for itself.

This chapter has been included because the mistaken claims of this group have been used as support for the suppression of the analysis I discuss in this book. A leading journal of mathematical statistics rather than of theoretical physics has quite rightly relied on the review of an eminent proponent of this mistake to decline the publication of related materials for professional examination. Thus, I refute it here, exposing the misunderstanding it entails.

Chapter 6 digresses from examination of Bell's inequality itself. Rather, it exposes several misunderstandings which are consequential for its long-imagined defiance. Particular confusions include the supposition that complex amplitude functions are peculiar to the unusual specification of quantum probabilities. In fact, they are representative of probability distributions assessed for any type of observations whatsoever. Further, the widespread allusion to the so-called "entanglement" of distant quantum particles derives simply from the symmetries assessed in the contexts in which they are observed. Such symmetries are central to the statistical analysis of scientific observations at every scale, most notably in agriculture and medicine, codified in the form of exchangeable distributions. These are only two of several misconceptions that are discussed. However, the overriding fundamental misconception underlying the assessment of quantum phenomena involves the nature of probability itself, and the role it plays in our understanding of physical experience.

Although there are many variations on the themes, the two main contending views of probability are the objective view and the subjective view. The former, largely supported within the physics community since the early twentieth century, is that randomness is a property of Nature, evident most notably at the scale of quantum experimentation. Indeed, the supposed defiance of Bell's inequality by quantum behaviour has been a motivating factor in support of this view. It underlies the traditional study of "statistical mechanics", proposed to govern the physical laws of mass random phenomena. Einstein was one among several physicists who could not subscribe to such an interpretation of probability, noted as he was for his adage that "the old one does not play dice".

In contrast, the subjective view of probability identifies it as the assessed measure of an individual's uncertain knowledge of observable occurrences of history, rather than as a generating feature of history itself. In agreement with this point of view, and as a proponent of its mathematical formalisation deriving from the work of Bruno de Finetti, I find common assessments of related matters by physicists to be riddled with confusion and error.

To be sure, there have been more recent developments in statistical mechanics following energetic arguments by Edwin Jaynes and a sizeable group of scholars touting maximum entropy procedures of statistical assessment based on a subjectivist outlook. Further, there has been a welcome introduction of subjective Bayesian perspectives on physical theory itself. These have been codified in the propositions of QBism, stimulated by the extensive writings of Christopher Fuchs and colleagues. The transformation of physical theory required to respond to these has yet to be completed.

Finally, Chapter 7 of this book addresses the original problem proposed by John Bell himself in assessing the challenge of Einstein to the "incompleteness" of the quantum theory. The challenge had appeared in a famous paper with Boris Podolsky and Nathan Rosen, known together as EPR. To an outsider, it appears that the community of theoretical physicists has been unduly protective of itself via self-adulation of its admittedly ingenious leading participants. It has been suggested at times that some should be excused from explaining their conclusions in public on account of their well-respected acumen. Such was never proposed by John Bell, who took care to address challenges to his work that arose during his lifetime, despite entertaining his own doubts. Nonetheless, he is appropriately revered for his lifetime of accomplishments. For this reason I leave to the final chapter a reassessment of the defiance of his inequality as he first formulated it in his initial response to EPR. Bemused as Bell was by the implications of his quantum research, I believe he would have been pleased to find embedded within it, too, the error of neglect exposed in the chapters of this book.

To conclude this Preface, I must express my dismay at the resistance of the community of theoretical physicists to an open discussion of the straightforward arguments that I present in these pages. The dissemination of these ideas in the ostensibly open arena of *arXiv* has been blocked by the trustees of that site, on the questionable grounds that they did "not have the expertise to review" them. It is a touted feature of this site that it disseminates unreviewed scientific and mathematical research. None of the characteristics they list as grounds for forbidding use of the site for exposition of scientific research pertains to the work I submitted to them. You can now assess it here for yourself.

Nonetheless, the central contents of this book have been published in one form or another in reviewed profession journals. To be sure, I had initially made attempts to publish Chapters 1 and 3 in pre-eminent journals of theoretical physics. Their editors refused to send my manuscript out for review on the grounds that the case for the defiance of Bell's inequality by quantum probabilities is closed, supposedly on the basis of overwhelming empirical evidence and unquestionable theoretical conclusions ... which I very pointedly dispute. My results have been published subsequently in reviewed journals: *Journal of Modern Physics, Entropy,* and *Applied Mathematics.*

One leading journal proposed a special issue on the assessment of Bell's inequality, and I did engagingly submit versions of three of these chapters there too for consideration. Reviewers' reactions were mixed, but the academic editor concluded to deny publication to two of them. Rather than addressing any technical detail of my argument, the editor agreed with the offended reviewers, and waxed eloquently about the intrinsic absurdity of quantum behaviour which must be recognised by anyone who wishes to consider the matter, and the inevitable requirement to recognise that the moon does not exist when we are not looking at her. Enough said.

Although this book concerns an involved mathematical argument, I believe the subject matter is of general interest to the educated public who are interested in the foundations of science ... even those who lie only on the fringe of technical competence with respect to its fine detail. A Google search of Bell's inequality brings up 3.6 million entries in .45 seconds. Despite the volume of this material, I believe my exposition here of the errors that have infested this discussion are novel. I have made some effort to write in a style that is conducive to a generally educated reader, and I have included some discussion that will make the ideas accessible more widely, beyond the limited realm of advanced mathematics. Although full attention to the details requires an intensity of concentration that is not common in the lay community, the mathematics involved is not really intricate. I hope the gist of the content will be understood and appreciated, even if you do not grind out every detail. You may rest assured, the moon is still encircling us, whether you are looking at her or not. Smile. Moreover, the oceans will continue to tide in her gravitational force, whether or not you are fortunate to watch.

Unashamedly, much of the exposition is addressed specifically to mathematicians and mathematical physicists. Being more than casually acquainted with the results of probability theory and its application, my analysis is based on well-established mathematical method, which follows the constructive subjectivist outlook of Bruno de Finetti. I would expect criticism of my exposition here, which I welcome, to be based on a technical assessment of my arguments, rather than merely offended rage.

While I have mentioned those unwilling to tender an exposition of the serious and evident errors I have found in established research, I must express my great appreciation to my friends Mike Ulrey, who first introduced me to the intrigue regarding Bell's inequality, and to Duncan Foley. Both have been supportive of my investigations, and attentive to critical details throughout extensive exchanges. Neither should be presumed to agree with all I have written. The University of Canterbury Department of Mathematics and Statistics provided computing support over several years. Thanks to the helpful IT staff: Paul Brouwers, Steve Gourdie, and Alan Witt, and to a colleague in the same department, Rachael Tappenden. It was she who programmed the fine details of the passage of a 4-dimensional polytope through 3-dimensional space.

Special thanks also to Giuseppe Sanfilippo, University of Palermo, for supportive discussions throughout this journey, and for his programming advice.

Having chided those who would suppress the dissemination of the contents of this book to the public, I would like to laud and to thank the editors of Austin Macauley Publishers who have dared to publish this material, which has been disdained by academic publishers on advice of eminent reviewers. In particular, thanks to Walter Stephenson, the copy editor, who has been very helpful. In this regard too, I would like to thank Jane Gao, a managing editor of the *Journal of Modern Physics* for her decision to publish the first run of results I have presented here. She has always insisted that critical reviews must be supported by rational argument rather than merely by reference to popular acclaim.

Frank Lad

Otautahi/Christchurch

Aotearoa/New Zealand

2 March, 2024

Chapter Summaries

1. WHAT IS SO WRONG?   a mathematical error of neglect

Bell's inequality was designed to assess consequences of Einstein's principle of local realism for quantum phenomena. In a specific CHSH format, it pertains to a linear combination of four polarisation products on the *same pair* of photons in a *gedankenexperiment*. After introducing the inequality in this context, we find that the summands of the crucial quantity $s(\lambda)$ inhere four symmetric functional relations that have long been neglected in its analytic consideration. The expectation $E[s(\lambda)]$ is not the sum of four "marginal" expectations from a joint distribution, as is typically construed. Rather, it has four distinct representations as the sum of three expectations plus the expectation of a fourth function of these three. Analysis using Bruno de Finetti's "fundamental theorem of probability" yields not the inequality-defying-value of $2\sqrt{2}$ for $E(s)$, but merely a bounded interval, $(1.1213, 2]$. The 4-D polytope of cohering joint P_{++} probabilities for the four stipulated angle settings is exhibited passing through 3-D space. Aspect's "estimation" of the component quantum expectations was based on polarisation products from different photon pairs that do not have embedded within them the restrictive functional relations underlying the inequality. When these restrictions are embedded into his estimation procedure in a specific way, it yields an estimate of 1.7667. Although this is not and cannot be a *definitive* assessment based on quantum theory, this same number emerges intriguingly in alternative characterisations of the problem as well. These will be exhibited in Chapter 5. The gist of this Chapter 1 first appeared in the *Journal of Modern Physics*, 2021.

2. EXTENDING THE SETUP TO TWO PAIRS OF PARTICLES: a gedankenexperiment requiring more denken

An influential article of Greenberger, Horne, Shimony, and Zeilinger "found" a contradiction in the analysis of a four-dimensional version of Bell's problem. It pertains to magnetic spin observations on a quartet of electrons. My exposition here shows that they inserted it into their analysis themselves by presuming contradictory premises: that the linear combination of the angles $\phi_1 + \phi_2 - \phi_3 - \phi_4$ involved in their proposed parallel experiments on two pairs of electrons equals both π and 0 at the same time. Unaware of this fundamental error, they continued their analysis, and proclaimed the impossibility of formalising local realism in problems involving more than two dimensions. After identifying their oversight, I expose how the notation they use in their "derivation" is not sufficient to represent the entanglement in the double electron spin-pair problem they propose. This confounds their mistake. The situation they design actually motivates only an array of symmetries involved in their experiment. These errors are not matters of interpretation or philosophy. They are basic errors of mathematical logic. The gist of this chapter first appeared in the journal *Entropy*, 2020.

3. RESURRECTING THE PRINCIPLE OF LOCAL REALISM

In this chapter I formalise a mathematical structure for the supplementary variable explanation of the results of theoretical quantum physics pertinent to the Aspect/Bell problem. Aspect had thought that a "naive distribution" he devised could be so represented, but that the quantum theoretic distribution could not. In fact, the same structure could underlie *any* probabilistic assessment of quantum observations whatsoever. Analysis based on the structure presented gives further support to recognition that the purported defiance of Bell's inequality by quantum probabilities is mistaken. The gist of this chapter first appeared in the journal *Applied Mathematics*, 2022.

4. MORE HOOJUMS THAN BOOJUMS:
    quantum mysteries for no one

In his very influential essay, "Quantum mysteries for anyone", David Mermin pulls a sleight of hand by exemplifying one experiment (a real one) with his machine, and another one (a gedankenexperiment) in his assessment of a proposed signal-response behaviour. The details pertain to the observation of magnetic spins on a pair of electrons that each address a Stern-Gerlach magnet at three possible angles. This provides for nine possible angle pairings. The gedanken formulation has embedded within it the same type of symmetric functional relations as pertain to the Aspect/Bell problem. His error in ignoring them is displayed in both a simulation procedure and in an assessment of the quantum gedankenexperiment he would have us ignore, while we are enticed to admire the mysterious behaviour of his machine. The gist of this chapter first appeared in the *Journal of Modern Physics*, 2022.

5. SIMULATING MULTI-POLARISATION IN A GEDANKENEXPERIMENT: the irrelevance of empirical averages

Although it is impossible to construct a physical experiment to assess the defiance of Bell's inequality, a mistaken consideration of the inequality as pertaining to experimental averages of actual experimental results has motivated claims of its empirical verification. A simulated version of the experiment is displayed which puts paid to such claims, exhibiting sensible results that preserve the principle of local realism in the assessment of quantum phenomena. Published averages of sequential polarisation observations of distinct photons are irrelevant to Bell's inequality.

6. ON PROBABILITY AND QUANTUM PHYSICS

Having attended exclusively to technical matters in the initial chapters of this exposition, I now remark on some interpretative matters of quantum physics. The discussion does border on technical issues. After reflecting on twentieth-century fascinations of quantum theory with matters of probability, I present a terse display of the formal structure of mathematical expectation as it has been developed following leading work by Bruno de Finetti. This is followed by an exhortation to recognise fundamental differences between expectations and averages. This distinction is quite commonly blurred in quantum discussions, with misleading consequences. Further dissection of distinctions between joint, conditional, and marginal probabilities is found relevant to common interpretations of quantum theory that I challenge as unwarranted and misguided. Moreover, an exposition of the surprising universal inherence of complex amplitude structures within *all* probability distributions defies common understanding of the uniqueness of quantum probabilities in requiring such underlying specifications. Finally, I identify the *exchangeability* of probability assertions as the basis for claims by physicists to particle *entanglement* as an objective feature of quantum behaviour. This property formalises the judgement of symmetry which is found pertinent to observable phenomena at all scales of experience. Concluding comments concern the vacuous concept of causality which cloud our understanding of the living exploding universe.

7. WHERE IT ALL STARTED: back to John Bell himself

I insist that the research results of all contributors to public scientific investigation should be assessed with the same scrupulous standards, even those of our most respected colleagues. Perhaps surprisingly, this is not an attitude shared by all in the physics community. In this final chapter,

I reassess and defy the original analysis of the appropriately heralded John Bell himself, in his famous response to Einstein and the EPR paper. It turns out that the feature of neglected functional relations among component "observations" in a gedankenexperiment is the source of his acclaimed discovery of the quantum defiance of the inequality. In fact, an analysis that recognises these relations exhibits the SATISFACTION of his inequality in the very problem he investigated. I propose that he would have been pleased to learn of this.

R.I.P. John Bell (1928-1990).

Chapter 1

WHAT IS SO WRONG ?
a mathematical error of neglect

This book will answer this question with complete details, from the origins of the error through more than sixty years of its infecting discussions of quantum theory. This first chapter introduces the basic problem as it has been described by Alain Aspect (2002), as well as the error involved in its widely accepted resolution. His exposition was prepared for a memorial lecture some ten years after John Bell's death, and some twenty years after Aspect's influential empirical investigation of the situation. Although different notations have been used during the course of subsequent discussions to which we shall refer, we shall begin using Aspect's now classic notation. I shall review both his mathematical formulation of the problem and his proposed solution, exposing the error involved for all to see. I suspect that many readers will have learned of the "mysteries of quantum physics" from descriptive YouTube presentations, which report the defiance of the inequality without explaining precisely how it arises.

Claims that probabilistic specifications of quantum mechanics defy the mathematical prescription known as Bell's inequality are just plain wrong. This may be difficult to accept, depending on how wedded you are to the outlook that arises from them. You will not be alone. The eminent journal *Nature* (2015, **526**, 649-650) flamboyantly announced to its readership the "Death by experiment for local realism" as an introduction to its publication of experimental results achieved at the Technical University of Delft. These were proclaimed to have closed simultaneously all seven loopholes that had been suggested as possible explanations of the celebrated inequality violation. In the eyes of the professional physics community, the matter is now closed. Three important propagators of the error, whom we shall meet in these pages, have been awarded the Nobel Prize for physics in 2022. My claim is that the touted violation derives from a mathematical mistake, an error of neglect. Its recognition relies only on a basic understanding of functions of many variables and on standard features of applied linear algebra. This presentation is designed for any sophisticated reader not put off by equations per se, who has followed this issue at least at the level of popular description of scientific activity.

It is clear in his own writings that John Bell himself was puzzled by the implications of his inequality, though he defended them from critics in several expositions (1971, 1975, 1981). Nonetheless, he suspected that something was wrong with the understanding that the probabilities of quantum mechanics seem to defy its structure, and he expressed undying confidence that this error would be discovered in due time. If the quantum world behaves according to principles distinct from those pertinent to larger scales, the classical world, there must be some boundary scale at which the principles change. At what scale might this occur? He wrote in 1971 and republished in 1987 the following remarks:

> "A possibility is that we will find where the boundary is. More plausible to me is that we will find that there is no boundary. It is hard for me to envisage intelligible discourse about a world with no classical part — no base of given events, be they only mental events in a single consciousness, to be correlated. On the other hand, it is easy to imagine that the classical domain could be extended to cover the whole. The wave function would prove to be a provisional or incomplete description of the quantum mechanical part, of which an objective account would become possible. It is this possibility, of a homogeneous account of the world, which is for me the chief motivation of the study of the so-called 'hidden variable' possibility."

Nonetheless, Bell deferred to his apparent mathematical results and resigned himself to the understanding that "no local hidden-variable theory can reproduce all the experimental predictions of quantum mechanics." This conclusion has been accepted and embellished over subsequent years of acclaimed research. I am making a bold claim that I have found a mathematical error which has been elusive. I accept all probabilistic assertions supported by quantum theory, and I exhibit their implied support of the inequality bounds introduced by Bell. It is in misrepresenting the defiance of the inequality that the error arises.

I do not contest the experimental results of the Delft group, nor any of the related experimentation that has followed from the path-breaking initial work of Aspect's group. I do contest the inferences they are purported to support. In this chapter I shall first review the derivation of the inequality in the context to which it applies, featuring its relation to Einstein's principle of local realism. The review will focus on the CHSH form of the inequality to which Aspect's optical experimentation is considered to be relevant. This acronym derives from the names of Clauser, Horne, Shimony, and Holt, authors who had introduced the design of the experiment in an important article (1969). Identifying the neglected functional relations that are involved in a thought experiment on a single pair of photons, I will show why the claims to defiance of the inequality are mistaken, and how to derive the actual implications of quantum theory for the probabilities under consideration. Further, it will be shown why Aspect's computations (and all subsequent extensions) proposed to exhibit empirical confirmation of the inequality defiance are ill-considered, and how they ought to be adjusted. This demonstration relies on the computational mechanics of Bruno de Finetti's fundamental theorem of probability.

The optical context for Bell's inequality

Herein we review the setup of an optical variant of Bell's experiment, designed in CHSH form by Alain Aspect (1981, 1982, with colleagues). Bell's original discussions (1964, 1966) of the inequality violation were couched in terms of observations of spins of paired electrons. Although specific algebraic details differ for the two types of experimental situation, the conclusions reached would be identical. We will have recourse to examine quantum probabilities for both contexts in the course of this book.

An experiment is conducted on a pair of photons travelling in opposite directions along an axis, $\mathbf{z}$, from a common source. The direction one of the photons travels toward detector A on the left is directly opposite to the direction its paired photon travels toward detector B on the right: $\mathbf{z}_A = -\mathbf{z}_B$ are the directional vectors. At the end of their respective journeys, each of the paired photons engages polarising material that allows it either to pass through or to be deflected, and thus be captured or not on photographic plates (not pictured) behind the polarisers.

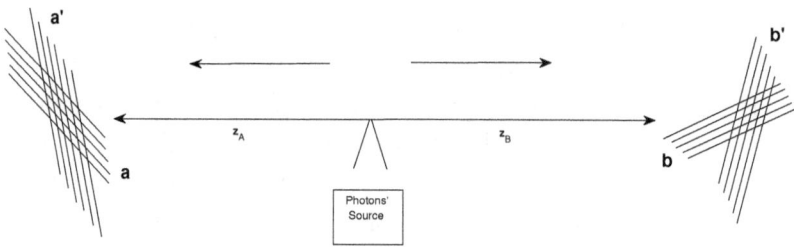

Figure 1: Polarising material is aligned at the detection stations of A and B, each with two possible choices of direction in the (x, y) dimension relative to the $\mathbf{z}$ direction of the incoming photon: direction $\mathbf{a}$ or $\mathbf{a}'$ at station A, and $\mathbf{b}$ or $\mathbf{b}'$ at station B.

The detection of a photon passing *through* the polariser is designated by denoting the numerical value of $A = +1$, while the detection of a photon as blocked is designated by the value of $A = -1$. The polariser addressed by photon A is directed at a variable angle $\mathbf{a}^*$ in the (x, y) plane perpendicular to $\mathbf{z}_A$. This polariser can be set in either of two directions in this plane, designated as $\mathbf{a}$ and $\mathbf{a}'$ in the experimental setups we shall consider. Similarly, the direction of the polariser met by the photon at station B can be set as either $\mathbf{b}$ or $\mathbf{b}'$ in its (x, y) plane. Depending on the specific pair of polarisation directions $\mathbf{a}^*$ and $\mathbf{b}^*$ chosen for any particular experiment, we shall observe the paired values of either $(A(\mathbf{a}), B(\mathbf{b}))$, or $(A(\mathbf{a}), B(\mathbf{b}'))$, or $(A(\mathbf{a}'), B(\mathbf{b}))$, or $(A(\mathbf{a}'), B(\mathbf{b}'))$. Since the observation values of the A and the B photon detections can each equal either $+1$ or -1 whatever the directional pairing might be, the chosen observation pair $(A(\mathbf{a}^*), B(\mathbf{b}^*))$ will equal one of the four possibilities $(+, +), (+, -), (-, +)$, or $(-, -)$, where we are suppressing here the needless numeric values of 1 in each designated pairing.

Experimental choices of the two polariser directions yield a specific *relative angle* between them at A and B in any given experiment. Using Aspect's notation that parentheses around a pair of directions denotes the relative angle between them, the experimental detection angle settings $(\mathbf{a}^*, \mathbf{z}_A)$ and $(\mathbf{b}^*, \mathbf{z}_B)$ imply the *relative* angle between polarisers at stations A and B in the (x, y) dimension as $(\mathbf{a}^*, \mathbf{b}^*)$. Bell's inequality is relevant to this context in which the two photon polarisation directions can be paired at any of four distinct relative angles, denoted by the parenthetic pairs $(\mathbf{a}, \mathbf{b})$, $(\mathbf{a}, \mathbf{b}')$, $(\mathbf{a}', \mathbf{b})$, or $(\mathbf{a}', \mathbf{b}')$, as displayed in Figure 2. Typically the choice of the polariser directions is made between the moment the photons leave their source and the moment they engage the polarisers. With photons travelling at speed of light, executing this is no mean feat.

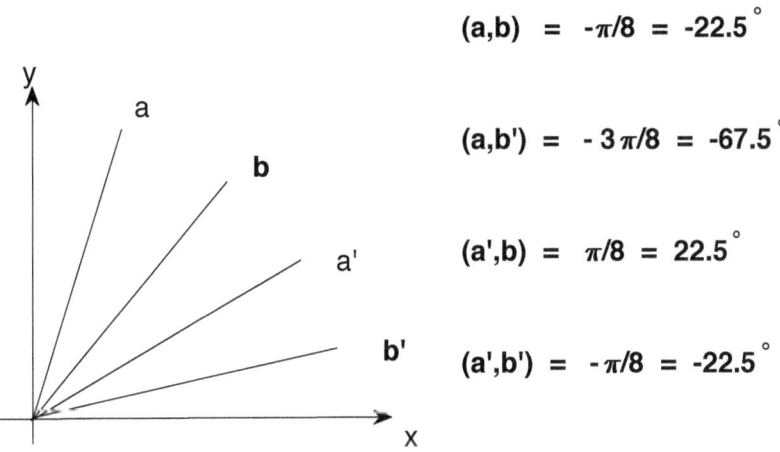

Figure 2: Directional vectors of the polarisation angle settings at the observation stations A and B, viewed in a common axis orientation. The specific relative angle size settings displayed are the extreme violation values, a feature to be discussed.

In order to view the *relative* polarisation angles at the two observation stations as shown, we would need mentally to swing the (x, y) plane around by $180°$ as it is viewed from the $\mathbf{z}_A$ direction by the photon approaching station A in Figure 1, and superimpose it on the (x, y) plane as viewed by the photon approaching station B from the $\mathbf{z}_B$ direction. In this manner we can understand the size and meaning of the relative angles *between* the various polariser directions $\mathbf{a}^*$ and $\mathbf{b}^*$ as the paired photons enter their detection stations.

The theory of quantum mechanics motivates the specification of probabilities for the four observable outcome possibilities of the polarisation experiment as depending on the *relative angle* $(\mathbf{a}^*, \mathbf{b}^*)$ between the direction vectors of the polarisers at stations A and B. For any such relative angle pairing, the probabilities specified by quantum theory for the four possible experimental observations $\{++, +-, -+, --\}$ are

$$
\begin{aligned}
P[(A(\mathbf{a}^*) = +1)(B(\mathbf{b}^*) = +1)] &= P[(A(\mathbf{a}^*) = -1)(B(\mathbf{b}^*) = -1)] \\
&= \tfrac{1}{2}\cos^2(\mathbf{a}^*, \mathbf{b}^*) ,
\end{aligned}
$$

and (1)

$$
\begin{aligned}
P[(A(\mathbf{a}^*) = +1)(B(\mathbf{b}^*) = -1)] &= P[(A(\mathbf{a}^*) = -1)(B(\mathbf{b}^*) = +1)] \\
&= \tfrac{1}{2}\sin^2(\mathbf{a}^*, \mathbf{b}^*) .
\end{aligned}
$$

For efficiency in what follows, we shall denote the four probabilities appearing in equations (1) by P_{++}, P_{--}, P_{+-}, and P_{-+} when the pertinent angle setting is evident.

These four probabilities surely sum to equal 1, because the sum of $cos^2 + sin^2$ of any angle equals 1. A few properties of the joint probability mass function (pmf) they compose should be noticed. Firstly, the four probabilities can be specified by the value of any one of them. The equations (1) stipulate that no matter what the relative angle $(\mathbf{a}^*, \mathbf{b}^*)$ may be, the values of $P_{++} = P_{--}$, and $P_{-+} = P_{+-}$. Since the four probabilities do sum to 1, the specification of P_{++} as the value p, for example, implies that the pmf vector $[P_{++}, P_{--}, P_{+-}, P_{-+}]$ would be $[p, p, (1-2p)/2, (1-2p)/2]$. Moreover, an important feature of these equalities, which we shall discuss more extensively in Chapter 6, is that the polarisation probabilities are symmetric, or "exchangeable". This is codified be the equality of the probabilities P_{+-} and P_{-+}.

Still another feature of this quantum distribution is that the probabilities for the paired detection outcomes depend only on the *product* of the two measurements. For both outcomes $++$ and $--$ yield a product of $+1$ and both outcomes $+-$ and $-+$ yield a product of -1. Thus, the QM-motivated probabilities for the experimental values of the polarisation *product* $A(\mathbf{a}^*)B(\mathbf{b}^*)$ are specified by

$$
P[A(\mathbf{a}^*)B(\mathbf{b}^*) = +1] = \cos^2(\mathbf{a}^*, \mathbf{b}^*)
$$

and

$$
P[A(\mathbf{a}^*)B(\mathbf{b}^*) = -1] = \sin^2(\mathbf{a}^*, \mathbf{b}^*).
$$

As will be important to recognise in what follows, the expected value of this distribution for the detection product is

$$E[A(\mathbf{a}^*)B(\mathbf{b}^*)] = (+1)cos^2(\mathbf{a}^*, \mathbf{b}^*) + (-1)sin^2(\mathbf{a}^*, \mathbf{b}^*)$$
$$= cos\, 2(\mathbf{a}^*, \mathbf{b}^*) \tag{2}$$

according to standard double angle formulas. It is worthwhile reminding right here that "the expected value of a probability distribution" is the "first moment" of the distribution. Geometrically, it is the point of balance of the probability mass function weights when they are positioned in space at the places where the possible observations to which they pertain might occur. It is a property of a probability distribution for the outcome of a specific single observable variable. It is not an average. Just saying.

A peculiarity of equation (2), which will be useful far down the road in this explication, is that the expectation value $E[A(\mathbf{a}^*)B(\mathbf{b}^*)]$ can also be represented as

$$E[A(\mathbf{a}^*)B(\mathbf{b}^*)] = 2\,cos^2(\mathbf{a}^*, \mathbf{b}^*) - 1 = 4\,P_{++}(\mathbf{a}^*, \mathbf{b}^*) - 1, \tag{3}$$

since the value of $sin^2(\mathbf{a}^*, \mathbf{b}^*)$ appearing in the final line of equation (2) can also be written as $1 - cos^2(\mathbf{a}^*, \mathbf{b}^*)$. Thus, the entire quantum distribution for the four possible polarisation observation pairs is also representable by the expectation of the polarisation product. Enough of this for now.

As conclusion to this discussion of the full joint distribution for the paired polarisation observations, the marginal probabilities that the detection observation of the photon equals $+1$ at either angle $\mathbf{a}^*$ or $\mathbf{b}^*$ is equal to $\frac{1}{2}$, no matter what the relative angle $(\mathbf{a}^*, \mathbf{b}^*)$ may be. For the standard margining equation for the result of a paired experiment yields

$$P[(A(\mathbf{a}^*) = +1) =$$
$$P[(A(\mathbf{a}^*) = +1)(B(\mathbf{b}^*) = +1)] + P[(A(\mathbf{a}^*) = +1)(B(\mathbf{b}^*) = -1)]$$
$$= \tfrac{1}{2}\,cos^2(\mathbf{a}^*, \mathbf{b}^*) + \tfrac{1}{2}\,sin^2(\mathbf{a}^*, \mathbf{b}^*) = \tfrac{1}{2}. \tag{4}$$

This result codifies a touted feature of physical processes at quantum scales of magnitude, that the photon behaviours of particle pairs are understood to be *entangled*. Since the probability for the joint photon behaviour $P[(A(\mathbf{a}^*) = +1)(B(\mathbf{b}^*) = +1)]$ does not factor into the product of their marginal probabilities $P[(A(\mathbf{a}^*) = +1)]$ and $P[(B(\mathbf{b}^*) = +1)]$, which are identical, the conditional distribution for either one of these events depends on the context of the conditioning behaviour:

$$P[(A(\mathbf{a}^*) = +1)|(B(\mathbf{b}^*) = +1)] \ = \ cos^2(\mathbf{a}^*, \mathbf{b}^*) \tag{5}$$

$$\neq \ P[(A(\mathbf{a}^*) = +1) \ = \ \tfrac{1}{2}$$

and

$$P[(A(\mathbf{a}^*) = +1)|(B(\mathbf{b}^*) = -1)] \ = \ sin^2(\mathbf{a}^*, \mathbf{b}^*),$$

which is different still.

We have concluded what we need to say at the moment about the prescriptions of quantum theory relevant to physical quantum behaviour of a single pair of prepared photons. Before proceeding to the specification of Bell's inequality, we need to address what quantum theory professes *not* to say.

The uncertainty principle:
what quantum theory disavows

Made famous as what is called "Heisenberg's uncertainty principle", the theory of quantum mechanics explicitly disavows claims to what might happen in physical situations that are impossible to instantiate. There is nothing very unusual about this. Here is an example relevant to the "classical scale" of everyday observation.

A dairy farmer may choose to treat a milking cow, beginning to-day, with injections of bovine somatatropine (BST), or may choose not to use such treatment. However, one cannot follow both pro-grammes on the same dairy cow. One could treat some cows with BST and some other cows without BST, but one cannot both treat and not treat the same cow with BST. Now what will be the daily weight of the milk yield from this cow during ensuing weeks? No one knows for sure ... under either condition. Neither the diary herder, nor you, nor I, nor anyone. Of course, you could well as-sess your *conditional* probability distributions for the weight of the cow's milk yield conditional on following each of the two treatment strategies. However, a distribution for the joint yields from fol-lowing both of these strategies on the same cow (which would be impossible) is meaningless. This is just common sense. Of course anyone is welcome to think about it. Thinking never hurt anyone.

There are well-known statistical procedures for studying the yields from alternative treatment strategies, applying one strategy to some cows and the other strategy to other cows whose initial conditions you regard symmetrically (exchangeably). This is a classic problem of agricultural statistics codified among problems in the *design of experiments*, which do not concern us here just now.

I present these considerations because the formulation of Bell's inequality involves this very same issue. We have identified a physical experiment on a pair of photons, polarising the two of them at an array of possible exclusive angle pairings, either $(\mathbf{a}, \mathbf{b})$, $(\mathbf{a}, \mathbf{b}')$, $(\mathbf{a}', \mathbf{b})$, or $(\mathbf{a}', \mathbf{b}')$. We could perform any of these experiments on a specific pair of photons at any one of these angle pairings. But we cannot perform all four experiments *on the same pair of photons*. The theory of quantum mechanics recognises this fact explicitly and loudly! Not only do the experimental physicists recognise that this cannot be done, but the theoretical algebraic mechanism that is used to identify the quantum probabilities we have specified in equations (1) embed this impossibility into its protocol. I will state how this recognition is embedded into quantum theory, without deriving its application here in complete detail.

Relevant to the polarisation product detection in the case we are formalising, the theory of quantum mechanics characterises the situation of the photons in a quantum experiment in terms of a two-dimensional vector which resides in several possible "states" of possibility. Let's call the state vector $\mathbf{s} = (s_1, s_2)^T$. We cannot observe what state the photon pair is in without performing a measurement. The measurement process is characterised algebraically by a matrix, call it H, which operates on the state vector by multiplication. The form of this matrix determines the observation values that might arise when the measurement is performed. In our case it specifies that the polarisation product value we observe might equal only $+1$ or -1, no matter what might be the paired polariser angles chosen in the experiment. When this matrix multiplies the state vector in the algebraic form of $H\mathbf{s}$, the result of the product yields a pair of probabilities specified for these two possible observation values. In our polarisation experiment on the photon pair observed at $A(\mathbf{a}^*, \mathbf{b}^*)$ and $B(\mathbf{a}^*, \mathbf{b}^*)$, these are the probabilities specified in our equation (1). (These two probabilities have been split in half there, to account for the fact that both of the paired polarisation results $++$ and $--$ yield a product of $+1$, and both of the paired polarisation results $+-$ and $-+$ yield a product of -1.)

At any rate, this is the mathematical formalism by which the probabilistic specifications of quantum theory are derived for the various possible results of a quantum experiment. Each array of paired observation possibilities in our optical experiment is characterised by its own matrix H. Since we have four possible experimental designs under consideration, codified by the paired angle settings $(\mathbf{a}, \mathbf{b})$, $(\mathbf{a}, \mathbf{b}')$, $(\mathbf{a}', \mathbf{b})$, and $(\mathbf{a}', \mathbf{b}')$, there are four distinct matrices, denoted by $H_{(\mathbf{a},\mathbf{b})}$, $H_{(\mathbf{a},\mathbf{b}')}$, $H_{(\mathbf{a}',\mathbf{b})}$, and $H_{(\mathbf{a}',\mathbf{b}')}$ which codify our experimental measurement possibilities for the selected polarisation angles.

Now, is it possible to perform *two* observational measurements on a quantum experimental situation? The answer is, "in some cases yes, and in some cases no!" Happily, there is a very simple way to determine whether two distinct measurements can be performed on the same situation of the state of the photons. Algebraically, a measurement codified by a matrix H is compatible with a simultaneous second measurement G on the same experimental situation if and only if the product of the two matrices commutes! That is to say, if and only if the products of the matrices are identical no matter which be the order of the multiplication: $HG = GH$. To check whether this is true or not is a simple matter of performing the algebra of multiplying the matrices.

An example of commuting operators has already been broached, without my having mentioned it. There is an operator matrix $H_{\mathbf{a}}$ that codifies the detection observation of the photon at station A with polarisation direction $\mathbf{a}$, and another which codifies a detection observation at station B with polarisation direction $\mathbf{b}$. Call it $H_{\mathbf{b}}$. Now it is a mere matter of mathematical derivation to find that the product of these two operator matrices does not depend of which one multiplies the other: that is, $H_{\mathbf{a}}H_{\mathbf{b}} = H_{\mathbf{b}}H_{\mathbf{a}}$. This lets us know formally that we can indeed measure the detection of the two photons at both of the stations A and B. That is why we denote this operator by $H_{(\mathbf{a},\mathbf{b})} \equiv H_{\mathbf{a}}H_{\mathbf{b}}$. We could easily have denoted it identically just as well by $H_{(\mathbf{b},\mathbf{a})} = H_{\mathbf{b}}H_{\mathbf{a}}$.

On the contrary, the result in the case of the paired photon experiments under consideration is that none of the four $H_{(\mathbf{a}^*,\mathbf{b}^*)}$ matrices commute! That is, for example, $H_{(\mathbf{a},\mathbf{b})}H_{(\mathbf{a},\mathbf{b}')} \neq H_{(\mathbf{a},\mathbf{b}')}H_{(\mathbf{a},\mathbf{b})}$. All this is to say that the technical manipulations of mathematical quantum theory instantiate formally just what we knew to begin with ... that we cannot simultaneously perform the measurement observation of the polarisation products at *both* angle settings $(\mathbf{a}, \mathbf{b})$ and $(\mathbf{a}, \mathbf{b}')$ on the *same pair of photons*.

Well, who would want to? We shall now find out.

"Local realism" and its relevance to Bell

A feature crucial to the touted violation of Bell's inequality is that it pertains to experimental results supposedly conducted with a *single photon pair* at all four angle settings. Sounds unusual? When the probabilistic pronouncements of quantum theory were formalised, Einstein among others was puzzled by the fact that the conditional probability for the outcome of the experiment at station A depends on both the angle at which the experiment is conducted at station B and on the outcome of that experiment. This matter is codified by the conditional probabilities we have seen in equations (5). This entanglement of seemingly unrelated distant physical processes was deemed to be a matter of "spooky action at a distance".

Well, Einstein proposed a solution to this enigma, positing that there must be some other factors relevant to what might be happening at the polariser stations A and B that would account for the photon detections found to arise. As yet unspecified in the theory, he considered such factors to identify unknown values of "supplementary variables". It was proposed that the probabilities inherent in the results of quantum theory must be representations of scientific uncertainty about the action of these other variables on the two photons at their respective stations. This was his way of accounting for the otherwise spooky action at a distance.

However, there was one aspect of the matter upon which Einstein wanted to insist: this was termed "the principle of local realism". Fair enough, quantum theory does stipulate the probability for the photon detection at angle setting $\mathbf{a}$ as depending on whether the polariser direction at B is set at $\mathbf{b}$ or $\mathbf{b}'$ and on what happens there. However, in any specific instance of the joint experiment at a relative polarisation angle $(\mathbf{a}, \mathbf{b})$, if the measurement observation at A happened to equal $A(\mathbf{a}, \mathbf{b}) = +1$, say, then in this instance the measurement at A would have to be the same no matter whether the direction setting at station B were $\mathbf{b}$ or $\mathbf{b}'$. That is to say, if the polarisation observation $A(\mathbf{a}) = +1$ in a particular experiment on a pair of photons measure in the paired angle design $(\mathbf{a}, \mathbf{b})$, then the value of $A(\mathbf{a})$ would also have to equal $+1$ in a companion experiment on the same pair of photons when the polarisation directions would be set in the angle pairing $(\mathbf{a}, \mathbf{b}')$.

Actually, in a way we have already deferred to such an understanding. We have been casually denoting the photon detection value at station A merely by $A(\mathbf{a})$ rather than denoting it by $A(\mathbf{a}, \mathbf{b})$, even before we have now introduced consideration of this principle of local realism. In the context of locality, the importance of such simplification of the notation was stressed by Aspect (2002) in his Bell memorial lecture. In fact, there was no need to denote the paired direction at B in our notation earlier, because we can only do the experiment on a specific photon at one specific possibility pair determining the angle pairing $(\mathbf{a}^*, \mathbf{b}^*)$. So we have merely denoted the measurements as $A(\mathbf{a})$ and $B(\mathbf{b})$, or $B(\mathbf{b}')$. Nonetheless, the QM *probabilities* of equations (1) stipulate that each of the paired results of the experiments does depends jointly on the *relative* angle between the two polarisation directions. If these probabilities were to constitute properties of the quantum system rather than properties of our uncertainty about it, this would amount to Einstein's designated "spooky action at a distance". How could the photon entering station A at polariser direction $\mathbf{a}$ know whether the paired photon is entering station B at polariser direction $\mathbf{b}$ or $\mathbf{b}'$?

Despite our notational deference, we should now recognise explicitly and declare loudly that this principle of local realism is based upon a claim that lies outside the bounds of matters addressed by the theory of quantum physics. For, as we have noted, it is impossible to make a measurement of both the photon detection product $A(\mathbf{a})B(\mathbf{b})$ *and* the product $A(\mathbf{a})B(\mathbf{b}')$ *on the same pair of photons*. So quantum theory explicitly disavows addressing this matter directly.

We are ready to conclude this Section by proposing an experimental measurement that lies at the heart of Bell's inequality. We are not yet ready to assess it, nor to explain its relevance to the principle of local realism, but we shall merely air it now for viewing. It is considered to be the result of a peculiar *gedankenexperiment*.

Consider a pair of photons to be ejected toward stations A and B at which the pair of polarisers can be directed in any of the four relative angles we have described. According to the detection of whether the photons pass through the polarisers or are deflected by them, Bell's inequality pertains to an experimental quantity

12

defined by the equation

$$s \equiv A(\mathbf{a}) B(\mathbf{b}) - A(\mathbf{a}) B(\mathbf{b}') + A(\mathbf{a}') B(\mathbf{b}) + A(\mathbf{a}') B(\mathbf{b}') \ . \quad (6)$$

Mathematically, we would describe this quantity s as a linear combination of four polarisation detection products. Any one of the four terms that determine the value of s could be observed in an experiment on a pair of prepared photons. Before we explain why this quantity is of interest, we should recognise forthrightly that we *can* observe the value of this quantity s if we are to conduct four component experiments on four *distinct* pairs of photons, each ejected toward stations A and B with the polarisers directed at a different relative angle pairing. However, we *cannot* observe the value of s if it were meant to pertain to all four experiments being conducted on the *same* pair of photons. It just cannot be done, and quantum theory is very explicit about having nothing directly to say about its value. If we are to consider the value of s in such an experimental design, it could only be as the result of a "thought experiment". Enough said for now.

Why would we even be interested by such a "gedankenexperiment" as its perpetrators called it, and what does the supposition of "hidden variables" have to do with the matter?

Einstein's proposal of hidden variables

Puzzled by the status of probabilistic conclusions of quantum theory which he helped to formulate, Einstein wondered what could be the meaning of these "probabilities" involved in its prescriptions. Others were proclaiming that the experimental and theoretical discoveries of QM support the view that at its fundamental level of particulate matter, the behaviour of Nature is random, and that quantum theory had identified its probabilistic structure. Convinced that "he (the old one) does not play dice with the universe", Einstein formulated another proposal: that the analysis of quantum theory is incomplete. There must be some other matters involved in quantum-level experimentation that we do not know about; and these other unknown "supplementary" variables would conceivably be distinguishable in the observable outcome of any particular experimental result, if only we knew how to distinguish their measurable states. The probabilities of quantum theoretical specifications must formalise our uncertain knowledge of the situation, our uncertainty distribution concerning the conceivable instantiation of these hidden variables in any particular experimental setting.

A necessity arises to formulate a gedankenexperiment in order to assess the matter. A paper by Einstein, Podolsky, and Rosen (1935) presented their argument which became known subsequently as the EPR proposal. It stimulated a fury of healthy discussion and argument that I shall not summarise here. Well documented both in the professional journals of physics and in literature of popular science, the discussion has featured considerations of the collapse of a quantum system when subject to observation that disturbs it, the apparent non-locality of quantum processes, and even esoteric formulations of the "many-worlds" view of quantum theory. What matters for my presentation here is that Einstein's views were widely relegated as a quirky peculiar sideline, and the recognition of randomness as a fundamental feature of quantum activity came to the forefront of theoretical physics.

Enter John Bell. Interested in a reconsideration of Einstein's view, he began his research with an idea to re-establish its validity as a contending interpretation of what we know. However, he was surprised to find this programme at an impasse when he discovered that if the principle of local realism were valid, the probabilistic specifications of quantum theory which we have described above seem to defy a simple requirement of mathematical probabilities. In the context of a hidden variables interpretation of the matter, this seemed to require that the principle of local realism must be rejected. Reported in a pair of articles (Bell, 1964, 1966), these results too stimulated an exacerbation of the fury which has continued even beyond the 2015 publication in *Nature* announcing their apparently definitive substantiation.

The specification of Bell's inequality can take many forms. The context in which it is addressed in the remainder of this chapter was presented in an article by Clauser, Horne, Shimony, and Holt (1969), commonly referred to as the CHSH formulation. This was the form that attracted still another principal investigator in this story, Alain Aspect. A young experimentalist, he wondered how such a monumental result of quantum physics could pertain only to a thought experiment, devoid of actual physical experimental confirmation. He proposed to devise and implement an experimental method that could confirm or deny the defiance of Bell's inequality.

My assessment of his empirical work follows directly from his explanation of the situation (Aspect, 2002) reported to a conference organised to memorialise Bell's work. My notation is largely the same as Aspect's. I adjust only the notation for expectation of a random variable to the standard mathematical notation of $E(X)$, replacing his notation of $<X>$ which has become standard in mathematical physics in the context of bra-ket notation which I avoid. Here is how it works.

Explicit construction of s with hidden variables

Hidden variables theory proposes that the quantity s which we have introduced in equation (6) should be considered to derive from a physical function of unobserved and unknown hidden variables, whose values might be codified by the vector λ, viz.,

$$
\begin{aligned}
s(\lambda) \equiv\ & A(\lambda, \mathbf{a})\, B(\lambda, \mathbf{b}) \ -\ A(\lambda, \mathbf{a})\, B(\lambda, \mathbf{b}') \\
& +\ A(\lambda, \mathbf{a}')\, B(\lambda, \mathbf{b}) \ +\ A(\lambda, \mathbf{a}')\, B(\lambda, \mathbf{b}') ,
\end{aligned} \qquad (7)
$$

for some unknown value of $\lambda \in \Lambda$. The variable designated by λ here could be a vector of any number of components identifying unknown features of the experimental setup that are relevant to the outcome of the experiment in any specific instantiation. The set designated by Λ is meant to represent the space of possible values of these hidden variables. The status of these variables in the context of any particular experiment is presumed to depend only on the state of the photon pair and its surrounds, independent of the angle setting $(\mathbf{a}^*, \mathbf{b}^*)$ at which the polarisers are directed. According to the functional outlook underlying physical theory, if we could only know the values of these hidden variables at the time of any experimental run, and have a complete theoretical understanding of their relevance to the polarisation behaviour of the photon pair, then we would know what the values of the polarisation incidence detection of the photon pair would be at any one or all of the possible angle settings.

Now the personalist subjective theory of probability (apparently subscribed to by Einstein, and surely by Bruno de Finetti and by me) specifies that any individual's uncertain knowledge of the values of observable but unknown quantities could be representable by a probability density over its space of possibilities. Aspect denotes such a density in this situation by $\rho(\lambda)$. For any proponent of the quantum probabilities, it might well be presumed to be "rotationally invariant" over the full array of angles at which the photons may be fluttering toward the polarisers. That is to say, the probabilities for the possible values of the supplementary variables do not depend on the angular direction in (x, y) dimensions of the photons along $\mathbf{z}$ axes heading toward stations A and B.

Since we avowedly have no idea of what these hidden variables may be, much less what their numerical values might be relevant to any specific experimental run, we can only ponder the "expected value" of $s(\lambda)$ with respect to the distribution specified by $\rho(\lambda)$. The feature of rotational invariance implies that this expectation would be the same, no matter what might be the rotational angle at which the photons flutter relative to their (x, y) plane detections. Let's write this expectation equation down:

$$
\begin{aligned}
E[s(\lambda)] \;=\; & E[A(\lambda, \mathbf{a})B(\lambda, \mathbf{b})] \;-\; E[A(\lambda, \mathbf{a})B(\lambda, \mathbf{b}')] \\
& + E[A(\lambda, \mathbf{a}')B(\lambda, \mathbf{b})] \;+\; E[A(\lambda, \mathbf{a}')B(\lambda, \mathbf{b}')] \quad (8)
\end{aligned}
$$

$$
= E[A(\mathbf{a})B(\mathbf{b})] - E[A(\mathbf{a})B(\mathbf{b}')] + E[A(\mathbf{a}')B(\mathbf{b})] + E[A(\mathbf{a}')B(\mathbf{b}')]
$$

as written by Aspect. There are three random variable terms in the initial expectation equation (8), these being A, B, and the lambda vector. Evaluation with respect to the rotationally invariant density $\rho(\lambda)$ yields the final combination of quantum probability expectations evaluating the polarisation behaviours.

Equation (8) follows directly from equation (7) because of the fact that a rule of probability says that the expectation of any linear combination of random quantities equals the same linear combination of their expectations. Fair enough. Fortunately, we have already reported in equation (5) that the probabilities of quantum theory identify the expected value of any polarisation product at the variable relative polarisation angle $(\mathbf{a}^*, \mathbf{b}^*)$ as $E[A(\mathbf{a}^*)B(\mathbf{b}^*)] = cos\, 2(\mathbf{a}^*, \mathbf{b}^*)$. So we are ready to proceed.

Finally, Bell's inequality

We have now arrived at a place we can state precisely what Bell's inequality says. There is just a little more detail to specify before we soon will have it. However, I should alert you that there is a small tic in the understanding of equation (8) to which we shall return, after we learn how the inequality is currently understood to be defied by quantum theory. But on the face of it, the validity of the equations (8) is plain as day.

Now re-examining equation (7), it is apparent that it can be factored into a simplified form:

$$s(\lambda) = A(\lambda, \mathbf{a}) \left[B(\lambda, \mathbf{b}) - B(\lambda, \mathbf{b}') \right] + A(\lambda, \mathbf{a}') \left[B(\lambda, \mathbf{b}) + B(\lambda, \mathbf{b}') \right],$$

or alternatively

$$(9)$$

$$= B(\lambda, \mathbf{b}) \left[A(\lambda, \mathbf{a}) + A(\lambda, \mathbf{a}') \right] - B(\lambda, \mathbf{b}') \left[A(\lambda, \mathbf{a}) - A(\lambda, \mathbf{a}') \right].$$

It is important to notice that once again, in performing this simple factorisation of the components $A(\lambda, \mathbf{a})$ and $A(\lambda, \mathbf{a}')$ in this second line, we have implicitly presumed the principle of local realism. For when we consider the first two summands of the first line, $A(\lambda, \mathbf{a}) B(\lambda, \mathbf{b})$, and $A(\lambda, \mathbf{a}) B(\lambda, \mathbf{b}')$, we should notice that the value of $A(\lambda, \mathbf{a})$ in that first term is evaluated in an experiment at which the paired polarisation angle is $(\mathbf{a}, \mathbf{b})$, whereas in the second term from which it is factored it is evaluated in an experiment at the relative polarisation angle $(\mathbf{a}, \mathbf{b}')$. It is the principle of local realism, extraneous to any claims of quantum theory, that provides that the observed value of $A(\lambda, \mathbf{a})$ must be identical in these two conditions on a single photon pair. It is impossible to instantiate them together. Only under the condition of local realism would we be able to factor this term out of the two expressions. The same goes for the factorisation of $A(\lambda, \mathbf{a}')$. This is not a source of any worry. I am merely mentioning this so that we are all aware of what is going on. The same feature of supposition is pertinent to the alternative factorisation of the terms $B(\lambda, \mathbf{b})$ and $B(\lambda, \mathbf{b}')$ in the third line from the terms of the first line.

Having arrived at this factorisation, it will now take just a little thought to recognise that if the value of the quantity s is supposed to be determined from a thought experiment on a single pair of photons, then the numerical value of s can equal only either $+2$ or -2. Of course, if we were to calculate the value of s from performing four component experiments with *four different pairs* of photons (something we can actually do), then the four component product values might each then equal either -1 or $+1$, so the value of s might equal any of $\{-4, -2, 0, +2, +4\}$. Notably, in such a case the factorisation we performed in equation (9) would not be permitted. For each of the observed detection products appearing in the first line would pertain to a different pair of photons whose multiplicands would be free to equal either $+1$ or -1 as prescribed by experiment. The same four possibilities would be accessible if the principle of local realism were *not* valid for a single photon pair. However, if the value of s is to be calculated from the results of a thought experiment on the same pair of photons, then its possibilities would be limited according to local realism merely to $\{-2, +2\}$. Here is how to recognise this.

Suppose the values of $B(\lambda, \mathbf{b})$ and $B(\lambda, \mathbf{b}')$ were both observed to equal $+1$. Then the first term in the factored form of the second line must equal $A(\lambda, \mathbf{a})\ [B(\lambda, \mathbf{b}) - B(\lambda, \mathbf{b}')] = 0$; and furthermore, the second term in the factored representation would then be $A(\lambda, \mathbf{a}')\ [B(\lambda, \mathbf{b}) + B(\lambda, \mathbf{b}')]$. The factor $A(\lambda, \mathbf{a}')$ equalling either $+1$ or -1 would then be multiplied by the factor $[B(\lambda, \mathbf{b}) + B(\lambda, \mathbf{b}')]$ which would equal the number $+2$. Thus, the value of s could equal only either -2 or $+2$. Alternatively, suppose that the values of $B(\lambda, \mathbf{b})$ and $B(\lambda, \mathbf{b}')$ are both observed to equal -1. Then by a similar argument the value of the first factored expression would again equal 0 and the second expression would equal either -1 or $+1$ multiplied now by -2. Again, the computed result of the value of s could equal only -2 or $+2$. The reader may confirm the same result for the possible values of s if the values of $B(\lambda, \mathbf{b})$ and $B(\lambda, \mathbf{b}')$ were observed to equal either -1 and $+1$ respectively, or $+1$ and -1 respectively.

The conclusion is indisputable. If the principle of local realism holds, then the value of s that would be instantiated as a result of a thought experiment on the same pair of photons in all four polarisation angle settings can equal only -2 or $+2$. Thus, the expected value $E(s)$ deriving from *any* coherent probability distribution over the four values of the component paired polarisation experiments would have to be a number between -2 and $+2$. Stated simply, without all the provisos explaining its content, Bell's inequality is the requirement that $-2 \leq E(s) \leq +2$.

Well, what do the probabilities of quantum theory imply for the value of $E(s)$? The answer universally presumed to be correct by proponents of the Bell violation is that when the design of the four experiments on a single pair of photons is constructed at a particular array of angle settings that we shall soon identify, then $E(s) = 2\sqrt{2} = 2.8284$ to four decimal places, a number that exceeds $+2$. (I shall show you why in the next paragraph.) Moreover, the experimental results of Aspect, as well as the more sophisticated experimentation of succeeding decades, are understood to corroborate this result to many decimal places. I will soon explain how this result is derived as well. However, I will insist on also showing you that not only is this theoretical derivation mistaken, but that the calculations used to corroborate this result from experimental evidence are incorrect. Nonetheless, there is nothing at all wrong with the experimental results, which are what they are. Nor is there anything wrong with Bell's inequality.

The purported violation of Bell's inequality

It turns out that Bell's inequality is *not* deemed to be defied at *every* four-plex of possible experimental angle settings, which we have characterised generically as $(\mathbf{a}, \mathbf{b})$, $(\mathbf{a}, \mathbf{b}')$, $(\mathbf{a}', \mathbf{b})$, or $(\mathbf{a}', \mathbf{b}')$. At some paired directional settings of the polarisers it seems not to be defied at all. Among other pairings at which it seems to be defied, it is apparently defied more strongly at some pairings than at others. Aspect opined that if we were to find experimental evidence of the defiance, we should try to find it at the angle pairings for which the theoretical defiance is the most extreme. It is a matter of simple calculus of extreme values to discover that the most extreme violation of the equality should occur at the angle settings $(\mathbf{a}, \mathbf{b}) = -\pi/8$, $(\mathbf{a}, \mathbf{b}') = -3\pi/8$, $(\mathbf{a}', \mathbf{b}) = \pi/8$, or $(\mathbf{a}', \mathbf{b}') = -\pi/8$. (The angle measurements are expressed here in terms of their polar representations. In terms of degrees, the angle $-\pi/8 = -22.5°$, while $-3\pi/8 = -67.5°$, and $+\pi/8 = +22.5°$.) You may wish to examine our Figure 2 and notice that the angles between the various polarisation directions we depicted there correspond to these relative angles. For the record, doubling these angles yields the values of $\pm\pi/4 = \pm45°$ and $-3\pi/4 = -135°$. And why does this matter?

Recall equation (8) and the ensuing sentences. Evaluating $E(s)$ according to this equation at the four angle settings just mentioned requires evaluating the summand component expectations. Each of them in the form $E[A(\mathbf{a}^*)B(\mathbf{b}^*)] = cos\,2(\mathbf{a}^*, \mathbf{b}^*)$, these would be

$$E[A(\lambda, \mathbf{a})B(\lambda, \mathbf{b})] \;=\; cos\,2(\mathbf{a}, \mathbf{b}) \;=\; cos(-\pi/4) \;=\; 1/\sqrt{2}\,,$$

$$E[A(\lambda, \mathbf{a})B(\lambda, \mathbf{b}')] \;=\; cos\,2(\mathbf{a}, \mathbf{b}') \;=\; cos(-3\pi/4) = -1/\sqrt{2}\,,$$

$$E[A(\lambda, \mathbf{a}')B(\lambda, \mathbf{b})] \;=\; cos\,2(\mathbf{a}', \mathbf{b}) \;=\; cos\,(\pi/4) \;=\; 1/\sqrt{2}\,, \text{ and}$$

$$E[A(\lambda, \mathbf{a}')B(\lambda, \mathbf{b}')] = cos\,2(\mathbf{a}', \mathbf{b}') \;=\; cos(-\pi/4) \;=\; 1/\sqrt{2}\,,$$

apparently yielding

$$E[s(\lambda)] \ = \ 1/\sqrt{2} - (-1/\sqrt{2}) + 1/\sqrt{2} + 1/\sqrt{2} \ = \ 4/\sqrt{2} \ = \ 2\sqrt{2}.$$

Voilà! The expected value of s apparently equals $2\sqrt{2} \approx 2.8284$, a real number outside of the interval $[-2, +2]$, defying Bell's inequality! What could be more simple, direct, and stunning?

Answer: . . . the truth!

OK, what is wrong, if anything?

The answer is found by thinking further about the result on which everyone agrees: the linear combination s can equal only -2 or $+2$ in the gedankenexperiment. Corollary to this should be recognition that if any three of its components sum to 3 or -1, then the fourth one *must* equal -1 in order for the value of s to satisfy the sum restriction. If the three would sum to -3 or $+1$, then the fourth one must equal $+1$. Technically, this constitutes a structure of symmetric functional relations among the four components of s which have long been ignored in the assessment of its expected value. The structure can be identified by constructing a matrix of all possible observation values that could result from performing the gedankenexperiment in CHSH form. Called "the realm matrix of observation possibilities" in the subjective theory of probability, we shall examine it in a partitioned form of its full extension as it pertains to every aspect of the problem, and then discuss it piece by piece.

I should mention that while the name "realm matrix of possibilities" has arisen within the operational subjective construction of probability theory, the matrix itself is merely a well-defined matrix of numbers that can be understood and appreciated by any experimentalist, no matter what your personal views about the foundations of probability might be. In the jargon of quantum physics it might be called the ensemble matrix of possible observation vectors.

A neglected functional dependence

In specifying the QM motivated expectation $E[s(\lambda)]$ as they do in our equation (8), Aspect/Bell fail to recognise a *symmetric functional dependence* among the values of the four proposed polarisation products composing $s(\lambda)$ as defined in equation (7), when it is meant to correspond to the result of the 4-ply thought-experiment on the same pair of photons. Perhaps surprisingly, the achieved values of any three *products* of the paired polarisation indicators imply a unique value for the fourth product. We now engage to substantiate this claim.

The realm matrix of experimental possibilities

Consider the realm matrix of all quantities relevant to the observations that might be made in the proposed 4-ply gedankenexperiment on a pair of photons under investigation. Unfortunately, its exhibition requires reduction of the type size. On the left side of the realm equation is written the name $\mathbf{R}(\mathbf{X})$, where $\mathbf{X}$ is a partitioned vector of names of every quantity that will be relevant to the outcome of the experiment, and to what quantum theory asserts about it. You will already recognise those in the first two partitioned blocks. On the right side of the realm equation appears a matrix whose columns exhaustively identify the values of these partitioned quantities that could possibly result from conducting the gedankenexperiment. We shall discuss them in turn.

$$
\mathbf{R}\begin{pmatrix} A(\mathbf{a}) \\ B(\mathbf{b}) \\ A(\mathbf{a}') \\ B(\mathbf{b}') \\ \ast\ast\ast\ast\ast \\ A(\mathbf{a})B(\mathbf{b}) \\ A(\mathbf{a})B(\mathbf{b}') \\ A(\mathbf{a}')B(\mathbf{b}) \\ A(\mathbf{a}')B(\mathbf{b}') \\ \ast\ast\ast\ast\ast \\ A(\mathbf{a}')\mathcal{B}(\mathbf{b}') \\ \ast\ast\ast\ast\ast \\ \Sigma_{/(\mathbf{a},\mathbf{b})} \\ \Sigma_{/(\mathbf{a},\mathbf{b}')} \\ \Sigma_{/(\mathbf{a}',\mathbf{b})} \\ \Sigma_{/(\mathbf{a}',\mathbf{b}')} \\ \ast\ast\ast\ast\ast \\ s(\lambda) \\ {}^{s}A/B(\mathbf{a}',\mathbf{b}') \\ 1 \end{pmatrix} =
$$

1	1	1	1	−1	−1	−1	−1	1	1	1	1	−1	−1	−1	−1
1	1	1	1	1	1	1	1	−1	−1	−1	−1	−1	−1	−1	−1
1	1	−1	−1	1	1	−1	−1	1	1	−1	−1	1	1	−1	−1
1	−1	1	−1	1	−1	1	−1	1	−1	1	−1	1	−1	1	−1
1	1	1	1	−1	−1	−1	−1	−1	−1	−1	−1	1	1	1	1
1	−1	1	−1	1	−1	1	−1	−1	1	−1	1	−1	1	−1	1
1	1	−1	−1	1	1	−1	−1	−1	−1	1	1	−1	−1	1	1
1	−1	−1	1	1	−1	−1	1	1	−1	−1	1	1	−1	−1	1
1	1	1	1	1	1	1	1	−1	−1	−1	−1	−1	−1	−1	−1
3	−1	−1	−1	1	1	−3	1	1	−3	1	1	−1	−1	−1	3
3	1	−1	1	1	−1	−3	−1	−1	−3	−1	1	1	−1	1	3
3	−1	1	1	−1	−1	−3	1	1	−3	−1	−1	1	1	−1	3
3	1	1	−1	−1	1	−3	−1	−1	−3	1	−1	−1	1	1	3
2	2	−2	2	2	−2	−2	−2	−2	−2	−2	2	2	−2	2	2
2	4	0	2	2	0	0	−2	−4	−2	−2	0	0	−2	2	0
1	1	1	1	1	1	1	1	1	1	1	1	1	1	1	1

The sixteen columns of four-dimensional vectors in the first partitioned block exhaustively list all the speculative 4×1 vectors of observation values that could possibly arise among the four experimental detections of photons at the four angles of polariser pairings. In order to observe the detection *products* at the four relative angles $A(\mathbf{a})B(\mathbf{b})$, $A(\mathbf{a})B(\mathbf{b}')$, $A(\mathbf{a}')B(\mathbf{b})$, and $A(\mathbf{a}')B(\mathbf{b}')$, we would surely have to observe each of the four *multiplicands* involved in their specification: $A(\mathbf{a})$, $B(\mathbf{b})$, $A(\mathbf{a}')$, and $B(\mathbf{b}')$. Since each of these observation values might equal only either −1 or +1, there

21

are sixteen possibilities of the 4-dimensional result of the 4-ply experiment. There are no presumptions made about these prospective quantity values: neither whether they "exist" or not prior to conducting the experiment, nor even whether they exist in any form after the experiment is conducted. We have merely made a list of what we could possibly observe if indeed we were capable of conducting the proposed gedankenexperiment on the same pair of photons. The observation vector would have to equal one of the 16 columns appearing in the top bank of the partitioned realm matrix.

Every other component quantity in the columns displayed in subsequent blocks of the realm matrix is computed via some function of the 4-D possibility vector above it in the column, for example the product values in the second partition. Notice once again that the "exhaustiveness" of this list presupposes the principle of local realism, specifying for example that the value of $A(\mathbf{a})$ identifying whether the photon passes through the polariser at A or not, would be the same no matter whether the polariser at which the paired photon engages station B is set at direction $\mathbf{b}$ or at $\mathbf{b}'$.

To begin the completion of the realm matrix, the second block of components identifies the four designated *products* of the paired polarisation indicators that yield the value of the quantity s as it is simply defined in equation (6). The first row of this second block, identifying the product $A(\mathbf{a})B(\mathbf{b})$, is the componentwise product of the first two rows of the first block. The second row of this block, identifying the product $A(\mathbf{a})B(\mathbf{b}')$, is the componentwise product of the first and fourth rows of the first block, and so on. This second block lists exhaustively in columns all the combinations of polarisation products that we could possibly observe in the conduct of our gedankenexperiment. Examine any one of these columns of products, checking that in fact the value of each product in that column is equal to the product of the corresponding multiplicands appearing in the column directly above it.

The first item to notice about this realm matrix is that, whereas the sixteen columns of the first block of polarisation observations are distinct, the second block contains only *eight* distinct column vectors. Columns 9 to 16 in block two of the realm matrix reproduce columns 1 to 8 in reverse order. Moreover, examining the first *three* rows of this second block more closely, it can be recognised that the first eight columns of these rows exhaust uniquely the simultaneous measurement possibilities for the three product

quantities they identify. These are the eight vectors of the cartesian product $\{+1, -1\}^3$, which are repeated in columns nine to sixteen in reverse order. Together, what these two observations mean
is that the fourth product quantity in this second block of vector
components is derivable as a function of the first three. What is
more, any one of the product quantities identified in block two is
determined by the same computational function of the other three!
This is what I meant earlier when alluding that the photon detection products in the gedankenexperiment have embedded within
them four symmetric functional relations. This can be seen by examining the columns of the *fourth* block of the matrix, which we
shall do shortly.

The third block of the realm matrix contains only a single row,
corresponding to a quantity we designate as $\mathcal{A}(\mathbf{a}')\mathcal{B}(\mathbf{b}')$. This quantity takes values only of ± 1, but it is logically independent of the
product quantities appearing in the first three rows of block two.
It takes the value of $+1$ in the first eight columns, and it takes the
value of -1 in the second eight columns in which the first three rows
of block two repeat themselves. This is the quantity Aspect/Bell
think they are assessing when they freely specify the quantum expectations for all four angle settings as they do, seemingly defying
Bell's inequality. We denote its name with calligraphic type to distinguish it from the actual polarisation product $A(\mathbf{a}')B(\mathbf{b}')$ whose
functional relation to the other three products we are now identifying. This singular component of the fourth partition block is
not an "Alice and Bob" observation quantity, but a rather peculiar "Aspect/Bell" imagined quantity. It is logically independent of
the first three "Alice and Bob" products. That is, whatever values
these products may be, the value of $\mathcal{A}(\mathbf{a}')\mathcal{B}(\mathbf{b}')$ may equal $+1$ in
the appropriate row among the first eight columns, or it may equal
-1 in the corresponding column among the second eight. However,
it does *not* represent the photon detection product $A(\mathbf{a}')B(\mathbf{b}')$ in
the four imagined experiments on a single photon pair.

Specifying the functional form via block four

Quantities in the fourth block of the realm matrix are designated
with the names $\Sigma_{/(\mathbf{a},\mathbf{b})}$, $\Sigma_{/(\mathbf{a},\mathbf{b}')}$, $\Sigma_{/(\mathbf{a}',\mathbf{b})}$, and $\Sigma_{/(\mathbf{a}',\mathbf{b}')}$. These quantities are defined by sums of column elements in *those rows of the
second block* that are *not marked* behind the slash in the subscript
notation. For examples,

$$\Sigma_{/(\mathbf{a},\mathbf{b})} \equiv A(\mathbf{a})B(\mathbf{b}') + A(\mathbf{a}')B(\mathbf{b}) + A(\mathbf{a}')B(\mathbf{b}'),$$

and $\quad \Sigma_{/(\mathbf{a},\mathbf{b}')} \equiv A(\mathbf{a})B(\mathbf{b}) + A(\mathbf{a}')B(\mathbf{b}) + A(\mathbf{a}')B(\mathbf{b}')\,.$

The quantities $\Sigma_{/(\mathbf{a}',\mathbf{b})}$, and $\Sigma_{/(\mathbf{a}',\mathbf{b}')}$ are defined similarly.

Next to notice is that the fourth row of the *second* matrix block, corresponding to $A(\mathbf{a}')B(\mathbf{b}')$, has an entry of 1 if and only if the fourth row of the *fourth* block, corresponding to $\Sigma_{/(\mathbf{a}',\mathbf{b}')}$, has an entry of -1 or $+3$ in the same column. When that entry is $+1$ or -3, the corresponding entry of the second block is -1. What this recognition does is to identify the functional relation of the fourth polarisation product to the first three polarisation products, viz.,

$$A(\mathbf{a}')B(\mathbf{b}') = \mathbf{G}[A(\mathbf{a})B(\mathbf{b}),\ A(\mathbf{a}')B(\mathbf{b}),\ A(\mathbf{a})B(\mathbf{b}')]$$

$$\equiv \left(\Sigma_{/(\mathbf{a}',\mathbf{b}')} = -1\,or\,+3\right) - \left(\Sigma_{/(\mathbf{a}',\mathbf{b}')} = +1\,or\,-3\right). \quad (10)$$

Here and throughout this book I am using notation in which parentheses surrounding a mathematical statement that might be true and might be false signifies the number 1 when the interior statement is true, and signifies 0 when it is false.

Some eyeball work is required to recognise functional relationship (10) by examining the final row of block two and of block four together. It may take even more concentration to recognise that the same or similar functional rule identifies the other three polarisation products as functions of the remaining three as well! The four product quantities $A(\mathbf{a}^*)B(\mathbf{b}^*))$ are related by four symmetric functional relationships, each of them being calculable via the same functional rule applied to the other three! This elusive recognition identifies the source of the Aspect/Bell error in assessing the QM-motivated expectation for $s(\lambda)$ as they do.

It is surely true that $E[s(\lambda)]$ equals a linear combination of four expectations of polarisation products, as specified in equation (8). Moreover, if the definition of $s(\lambda)$ in equation (6) were understood to represent the combination of observed products from experiments on four distinct pairs of photons, then the possible values of $s(\lambda)$ would include the integers $\{-4, -2, 0, 2, 4\}$; the expectation of each product $E[A(\mathbf{a}^*)B(\mathbf{b}^*)]$ would equal $-1/\sqrt{2}$ or $+1/\sqrt{2}$ as appropriate to the angle $(\mathbf{a}^*, \mathbf{b}^*)$; and $E[s(\lambda)]$ would equal $2\sqrt{2}$ as proposed by Aspect/Bell. This involves no violation of any probabilistic inequality at all, and there is no suggestion of mysterious activity of quantum mechanics.

However, when it is proposed that the paired polarisation experiments at the four angles are imagined pertaining to the same photon pair, then each of the products is restricted to equal a function value of the other three, as we identified explicitly for $A(\mathbf{a}')B(\mathbf{b}')$ via the function $\mathbf{G}[A(\mathbf{a})B(\mathbf{b}), A(\mathbf{a}')B(\mathbf{b}), A(\mathbf{a})B(\mathbf{b}')]$. In this context, Aspect's expected quantity would be representable equivalently by any of the following equations:

$$
\begin{aligned}
E[s(\lambda)] &= E[A(\mathbf{a})B(\mathbf{b})] - E[A(\mathbf{a})B(\mathbf{b}')] + E[A(\mathbf{a}')B(\mathbf{b})] \\
&\quad + E\{\mathbf{G}[A(\mathbf{a})B(\mathbf{b}), A(\mathbf{a})B(\mathbf{b}'), A(\mathbf{a}')B(\mathbf{b})]\} \\
&= E[A(\mathbf{a})B(\mathbf{b})] - E[A(\mathbf{a})B(\mathbf{b}')] + E[A(\mathbf{a}')B(\mathbf{b}')] \\
&\quad + E\{\mathbf{G}[A(\mathbf{a})B(\mathbf{b}), A(\mathbf{a})B(\mathbf{b}'), A(\mathbf{a}')B(\mathbf{b}')]\} \\
&= E[A(\mathbf{a})B(\mathbf{b})] + E[A(\mathbf{a}')B(\mathbf{b})] + E[A(\mathbf{a}')B(\mathbf{b}')] \\
&\quad - E\{\mathbf{G}[A(\mathbf{a})B(\mathbf{b}), A(\mathbf{a}')B(\mathbf{b}), A(\mathbf{a}')B(\mathbf{b}')]\} \\
&= -E[A(\mathbf{a})B(\mathbf{b}')] + E[A(\mathbf{a}')B(\mathbf{b})] + E[A(\mathbf{a}')B(\mathbf{b}')] \\
&\quad + E\{\mathbf{G}[A(\mathbf{a})B(\mathbf{b}'), A(\mathbf{a}')B(\mathbf{b}), A(\mathbf{a}')B(\mathbf{b}')]\} \quad (11)
\end{aligned}
$$

The symmetries involved in the setup would support an identical result in each case, and it would surely *not* yield $2\sqrt{2}$ at all. *This is the mathematical error of neglect to which the title of this current chapter alludes.* What might it yield?

The functional relation we have exposed in (10) is *not* linear. If it were, then the specification of an expectation for its arguments would imply the expectation value for the function value. As it is not, the specification of expectation values for the arguments only implies *bounds* on any cohering expectation value for the fourth. These numerical bounds can be computed using a theorem due to Bruno de Finetti, which he first presented at his famous lectures at the *Institut Henri Poincaré* in 1935. He named it only in his swansong text (de Finetti, 1974, 1975). It was first characterised in the form of a linear programming problem by Bruno and Giglio (1980), and has appeared in various forms in recent decades. Among them are presentations in dual form by Whittle (1970, 1971) using standard formalist notation and objectivist concepts. We shall review the content of de Finetti's theorem shortly, and then examine its relevance to assessing the expectation of $s(\lambda)$ motivated by considerations of quantum mechanics. We need first to air some further brief remarks about the final block of the realm matrix.

The final block of quantities and their realm

The first row of block five of the realm matrix merely identifies the values of $s(\lambda)$ associated with the polarisation observation possibilities enumerated in the columns of block one. Each component of this row is computed from the corresponding column of block two according to equation (6) which defines the quantity s. It is evident that every entry of this row is either -2 or $+2$. This verifies the argument we have made following the factorisation equation (9).

The second row of this block pertains to a quantity denoted as $s_{\mathcal{A}/\mathcal{B}(\mathbf{a}',\mathbf{b}')}$. Its value is defined similarly to equation (6), but its final summand is specified as the Aspect/Bell quantity $\mathcal{A}(\mathbf{a}')\mathcal{B}(\mathbf{b}')$ introduced in block three rather than the actual polarisation product quantity $A(\mathbf{a}')B(\mathbf{b}')$ which appears in the equation defining $s(\lambda)$. The realm of $s_{\mathcal{A}/\mathcal{B}(\mathbf{a}',\mathbf{b}')}$ can be seen to include the elements $\{-4, -2, 0, 2, 4\}$ whereas the realm of $s(\lambda)$ includes only $\{-2, 2\}$. The fact that the possibilities for $s_{\mathcal{A}/\mathcal{B}(\mathbf{a}',\mathbf{b}')}$ include both -4 and $+4$ is what makes it not surprising that the expectation of *this quantity* is $2\sqrt{2}$ as pronounced by proponents of the Aspect/Bell analysis.

The final row of block five is merely an accounting device, denoting that the "sure" quantity, the number 1, is equal to 1 no matter what the observed results of the four imagined optical experiments might be. Its relevance will become apparent when the need arises to apply de Finetti's fundamental theorem to quantum assertions.

It is time for a rest and an interlude. It is a mathematical interlude whose complete understanding relies only on your knowledge of some basic methods of linear algebra, currently introduced in secondary schools. If you would like a slow didactic introduction to the subject, my best suggestion is to look at Chapter 2.10 of my book, Lad (1996). You may even wish to start in Section 2.7. Another purely computational presentation appears in the article of Capotorti et al (2007, Section 4). I will make another attempt here in a brief format, merely to keep this current exposition self-contained. What does the fundamental theorem of probability say?

The fundamental theorem of probability

In brief, the fundamental theorem says that when you specify expectation values for any vector of quantities whatsoever, then the rules of probability provide numerical bounds on a cohering expectation for any other quantity you would like to assess. These can be computed from a linear programming routine. If the expectations you have specified are incoherent (meaning self-contradictory) among themselves, then the linear programming problems they motivate have no solution. This theorem is immediately relevant to our situation here in which we have identified quantum-theory-motivated expectations for any three of the four detection products that determine the value of s for the gedankenexperiment. We wish to find the bounds on the cohering expectation for the fourth detection product, restricted to equal a function value determined by these three. In brief, here is how the theorem works.

Suppose you have identified the expectations for N quantities, and you are wondering what you might assert as the expectation for another one, call it the $(N + 1)^{st}$. What you should do to evaluate the range of cohering possibilities for this expectation is firstly to construct the realm matrix of possible values for the vector of all $(N + 1)$ quantities. Let's denote the vector by $\mathbf{X}_{N+1} = (X_1, X_2, ..., X_N, X_{N+1})^T$, and call its realm matrix $\mathbf{R}(\mathbf{X}_{N+1})$. In general, it will look something like the realm matrix we have just constructed for various aspects of our gedankenexperiment. It will have $N+1$ rows, and some number K columns. Just as an example, the realm matrix we have already constructed happens to have 16 rows and $K = 16$ columns. (Mind you, we have not yet specified expectation for the first 15 components of the quantity vector to which this realm applies, but let's not let that deter us. I am merely suggesting here an example of a realm matrix that could be considered to have $(N+1)$ rows. Let's continue with the general abstract specification.)

Now any such vector of quantities can be expressed as the product of its realm matrix with a particular vector of events. These events are defined in terms of the finite possibilities for the quantity vector. The matrix equation, displayed in a form that partitions the final row, would look like this:

$$
\begin{pmatrix} X_1 \\ X_2 \\ X_3 \\ \cdot \\ \cdot \\ X_N \\ {*}{*}{*} \\ X_{N+1} \end{pmatrix}
=
\begin{pmatrix}
x_{1,1} & x_{1,2} & \cdot & \cdot & x_{1,K} \\
x_{2,1} & x_{2,2} & \cdot & \cdot & x_{2,K} \\
x_{3,1} & x_{3,2} & \cdot & \cdot & x_{3,K} \\
\cdot & \cdot & \cdot & \cdot & \cdot \\
\cdot & \cdot & \cdot & \cdot & \cdot \\
x_{N,1} & x_{N,2} & \cdot & \cdot & x_{N,K} \\
{*}{*}{*} & {*}{*}{*} & * & * & {*}{*}{*} \\
x_{(N+1),1} & x_{(N+1),2} & \cdot & \cdot & x_{(N+1),K}
\end{pmatrix}
\begin{pmatrix}
(\mathbf{X}_{N+1} = \mathbf{x}_{\cdot 1}) \\
(\mathbf{X}_{N+1} = \mathbf{x}_{\cdot 2}) \\
(\mathbf{X}_{N+1} = \mathbf{x}_{\cdot 3}) \\
\cdot \\
\cdot \\
(\mathbf{X}_{N+1} = \mathbf{x}_{\cdot (K-1)}) \\
(\mathbf{X}_{N+1} = \mathbf{x}_{\cdot K})
\end{pmatrix}
.
$$

On the left of this equation is the column vector of the quantity observations under consideration, with its final element partitioned.

To the right of the equality comes firstly the $(N + 1) \times K$ realm matrix whose K columns list all the possible columns of numbers that could possibly result as the observation vector. Again, the final row has been partitioned to appear on its own. The row will be designated in what follows as $\mathbf{R}(X_{N+1})$. (Notice this "X" is not bold, since the quantity X_{N+1} is merely the $(N+1)^{st}$ component of $\mathbf{X}_{N+1}$.) The K columns, each of which has $(N+1)$ components, correspond to vectors which can be denoted as $\mathbf{x}_{\cdot 1}, \mathbf{x}_{\cdot 2}, \mathbf{x}_{\cdot 3}, ..., \mathbf{x}_{\cdot(K-1)}$, and $\mathbf{x}_{\cdot K}$. (The initial subscripted dot denotes that this is a whole column of numbers. The number that follows the dot denotes which of the columns of the matrix it is we are talking about.) This matrix is multiplied by the final $K \times 1$ column vector of events that identify whether the quantity vector $\mathbf{X}_{N+1}$ turns out upon observation to be the first, the second, ..., the $(K - 1)^{st}$, or the Kth of these realm matrix columns. We denote this vector by $\mathbf{Q}(\mathbf{X}_{N+1})$, and call it "the partition vector generated by $\mathbf{X}_{N+1}$". One and only one of its component events will equal 1 and the rest will equal 0. But we do not know which of them is the 1, because we do not know which column of possibilities in the realm will represent the observed outcome of the vector of quantities $\mathbf{X}_{N+1}$.

We can represent this matrix equation more concisely and in a useful form by writing it in an abbreviated partitioned form:

$$\begin{pmatrix} \mathbf{X}_N \\ X_{N+1} \end{pmatrix} = \mathbf{R} \begin{pmatrix} \mathbf{X}_N \\ X_{N+1} \end{pmatrix} \mathbf{Q}(\mathbf{X}_{(N+1)}) \ .$$

The payoff from constructing this matrix structure is that now every row of this partitioned equation has on its left-hand side the unknown value of a quantity, X_i. On the right-hand side in that row appears a list of the possible values of that quantity in the context of possible values of the other quantities shown in that column as well. Each of the row possibilities is multiplied in a linear combination with the final column of events that specifies whether each of them is indeed observed as the value of this quantity, or not. Each row of this equation specifies how a different one of the quantities under consideration equals a linear combination of events.

We have heard of this before. The expectation of a linear combination of events equals the same linear combination of expectations for those events. These would be probabilities for the components of the $\mathbf{Q}$ vector if we could specify values for them. This tells us that we can evaluate an expectation operator on this partitioned equation to yield

$$E \left(\begin{array}{c} \mathbf{X}_N \\ X_{N+1} \end{array} \right) = \mathbf{R} \left(\begin{array}{c} \mathbf{X}_N \\ X_{N+1} \end{array} \right) P[\mathbf{Q}(\mathbf{X}_{(N+1)})] \ .$$

Well, we have not mentioned anything about probability specifications appearing in the vector $P[\mathbf{Q}(\mathbf{X}_{(N+1)})]$ on the right-hand side of this equality. The only restrictions on these probabilities are that they must be non-negative numbers that sum to 1, since the vector $\mathbf{Q}(\mathbf{X}_{(N+1)})$ constitutes a partition. We have mentioned only that *expectations* have been identified for the first N components of the vector on the left-hand side, $E(\mathbf{X}_N)$. Yet we can compute something important on the basis of this realisation. The linearity of this equation ensures that the implied value for the expectation of the final unspecified component $E(X_{N+1})$ must lie within a specific interval. It is computable as the

minimum and maximum values of $\mathbf{R}(X_{N+1})\mathbf{q}_K$

subject to the linear restrictions $\mathbf{R}(\mathbf{X}_N)\mathbf{q}_K = E(\mathbf{X}_N)$,

as is required of the expectations we have presumed to be specified, where the components of $\mathbf{q}_K$ must be non-negative and sum to 1. Such a computation is provided by the compilation of a linear programming problem.

The "solutions" to these linear programming problems are the vectors $\mathbf{q}_{min}$ and $\mathbf{q}_{max}$ that yield these minimum and maximum values for $E(X_{N+1})$ subject to these constraints. The final row vector identifying $E(X_{N+1})$ whose extreme values we seek is called "the objective function" of the problems. Its coefficients are the partitioned final row of the general realm matrix, identified as $\mathbf{R}(X_{N+1})$. Notice again that this X is not bold. It represents merely the final quantity in the column vector $\mathbf{X}_{N+1}$. The coefficient vector of the objective function is the final row vector of the full realm matrix.

Here are specific exemplary details that are appropriate to the gedankenexperiment we have been considering:

$$\mathbf{E} \left(\begin{array}{c} 1 \\ A(\mathbf{a})B(\mathbf{b}) \\ A(\mathbf{a})B(\mathbf{b}') \\ A(\mathbf{a}')B(\mathbf{b}) \\ A(\mathbf{a}')B(\mathbf{b}') \end{array} \right) = \left(\begin{array}{cccccccc} 1 & 1 & 1 & 1 & 1 & 1 & 1 & 1 \\ 1 & 1 & 1 & 1-1-1-1-1 \\ 1-1 & 1-1-1 & 1-1 & 1 \\ 1 & 1-1-1 & 1 & 1-1-1 \\ 1-1-1 & 1 & 1-1-1 & 1 \end{array} \right) \mathbf{q}_8 \ . \quad (12)$$

I have listed the order of the quantities in the vector at left to begin with the sure quantity, 1, which equals 1 no matter what happens in the gedankenexperiment. There follow the four summands of the CHSH quantity s, of which we have noticed that each one of them is restricted in the gedankenexperiment to equal a function value of the other three. That is why there are only eight columns in their realm matrix, as opposed to sixteen columns in the expansive realm matrix we have examined initially.

As to the components of the vector $\mathbf{q}_8$ on the far right-hand side of (12), notice that quantum theory says nothing at all about these, individually. Each of them should equal the probability that the 4-ply gedankenexperiment would produce results designated by a specific column of the realm matrix. However, these would involve the joint detection of polarisation products in four distinct measurements that are known to be incompatible. On account of the generalised uncertainty principle, quantum theory eschews specification of such probabilities. Nonetheless, for any individual polarisation product for a specific experimental design appearing on the left-hand side of the equation, quantum theory does specify its expectation value as either $1/\sqrt{2}$ or $-1/\sqrt{2}$, as we have recognised. Since these four products would not *all* be free to equal $+1$ or -1 at the same time due to the restriction on s to equal only either -2 or $+2$, we may assert expectation values for any three of them, and use linear programming computations to find the cohering bounds on the expectation of the fourth that would accompany them. This would yield bounds on the expectation of the Bell/Aspect CHSH quantity $E(s(\lambda))$ as well, identified in equations (11).

Quantum theoretic restrictions on valuation of $\mathbf{q}_8$

This is what we find. The columns of the matrix below display the computed results of the paired $\mathbf{q}_{min}$ and $\mathbf{q}_{max}$ vectors corresponding to four linear programming problems. Each of them determines a bound on an expected function value that appears in one of the four forms of the expectation value $E[s(\lambda)]$ displayed in equation (11). The first pair of columns, for example, identify extreme LP solution vectors for the objective function $E[A(\mathbf{a}', \mathbf{b}')]$ appearing as the fifth row of the matrix in equation (12), constrained by QM-specified values of the expectations of the first four rows. The second pair of columns identify extreme LP solutions when the fourth row of (12) specifies the objective function, constrained by QM specifications of expectations for rows $1, 2, 3$, and 5; and so on.

	q_{min} (a′, b′)	q_{max} (a′, b′)	q_{min} (a′, b)	q_{max} (a′, b)	q_{min} (a, b′)	q_{max} (a, b′)	q_{min} (a, b)	q_{max} (a, b)
q_1	0	.1464	0	.1464	.5607	.7803	0	.1464
q_2	.7803	.5607	0	.1464	.1464	0	0	.1464
q_3	.0732	0	.0732	0	0	.0732	0	0
q_4	0	.1464	.7803	.5607	.1464	0	0	.1464
q_5	0	.1464	0	.1464	.1464	0	.7803	.5607
q_6	.0732	0	0	0	0	.0732	.0732	0
q_7	.0732	0	.0732	0	0	0	.0732	0
q_8	0	0	.0732	0	0	.0732	.0732	0

Each of these column vectors resides in 8-dimensional space, providing an extreme coherent assessment of probabilities for the constituent event vector $\mathbf{Q}(\mathbf{X}_8)$, without specifying precise probabilities for any of them. In fact, quantum theory denies itself the capability of identifying such probabilities precisely. However, the results of the linear programming computations can and do specify extreme possibilities for what would cohere with what quantum theory can and does tell us. Any convex (linear) combination of these extremes would cohere with quantum theory as well. Thus, geometrically, the columns displayed constitute vertices of a polytope of quantum-theory-supported possibilities for $P[\mathbf{Q}(\mathbf{X}_8)]$. This polytope is called "the convex hull" of these vectors. The dimension of this hull is somewhat smaller. Although we have found eight vertices, the rank of the matrix of all of them is only four! That is, these eight-dimensional vectors all reside within a four-dimensional subspace of a unit-simplex.

Why is quantum theory not more specific in specifying the expectation of Bell's quantity $E(s)$? We shall delay this discussion until we have clarified what we have learned from these results of $\mathbf{q}_{min}(\mathbf{a}^*, \mathbf{b}^*)$ and $\mathbf{q}_{max}(\mathbf{a}^*, \mathbf{b}^*)$.

Implied bounds on expected detection products and on $E(s) \in (1.1213, 2]$

According to the prescription of equation (12), each of these *min* and *max* $\mathbf{q}_8$ vectors would identify a vertex of a polytope of cohering expectation vectors for the components of the CHSH quantity s. Followed at bottom by the implied expectation value $E(s)$, these vertices are constituted by the columns of the following display:

$E[A(\mathbf{a})B(\mathbf{b})]$	.7071	.7071	.7071	.7071	.7071	.7071	-1.0000	$-.1213$
$E[A(\mathbf{a})B(\mathbf{b}')]$	$-.7071$	$-.7071$	$-.7071$	$-.7071$	.1213	1.0000	$-.7071$	$-.7071$
$E[A(\mathbf{a}')B(\mathbf{b})]$	.7071	.7071	-1.0000	$-.1213$	.7071	.7071	.7071	.7071
$E[A(\mathbf{a}')B(\mathbf{b}')]$	-1.0000	$-.1213$	.7071	.7071	.7071	.7071	.7071	.7071
$E[s]$	1.1213	2.0000	1.1213	2.0000	2.0000	1.1213	1.1213	2.0000

In any of these columns appear three values of $E[A(\mathbf{a}^*)B(\mathbf{b}^*)]$ specifications supported by quantum theory, and a fourth value which is either a lower bound or upper bound on any cohering expectation for the fourth. (By the way, 0.7071 is the value of $1/\sqrt{2}$ to four decimal places.) At the bottom of the column is the value of $E(s)$ that would correspond to these four. The vectors of four $E[A(\mathbf{a}^*)B(\mathbf{b}^*)]$ values are vertices of a four-dimensional space of QM-supported expectation values of the gedankenexperiment. The value of $E(s)$ listed at bottom would be an extreme quantum-theory-permitting assessment of $E(s)$, Bell's quantity. All of their convex combinations lie within Bell's reputed bounds of $[-2, +2]$. There is more that can be said about this, but let us first address the question of why quantum theory leaves four dimensions of freedom unaccounted for in its prescriptions.

Why four free dimensions to specification of $E(s)$?

Let's just get down to it, without any prelude. Quantum theory specifies precise values for outcome probabilities of the photon pair detections at any choice of three angle settings of the gedankenexperiment. Consider the QM identifications of detection probabilities at the paired polarisation angles $(\mathbf{a}, \mathbf{b})$, $(\mathbf{a}, \mathbf{b}')$, and $(\mathbf{a}', \mathbf{b})$ which were employed as constraints in the first pair of LP problems. These have been displayed in our equations (1), and the corresponding expectations of the detection products have appeared in equation (2). However, if quantum theory were to specify a *complete* distribution for the outcome of this gedankenexperiment at all three of these angle pairings simultaneously, it would have to specify *eight* probabilities pertinent to the three. These would involve not only the three corresponding to detection events at each of the polarisation angles, but also jointly at any two of the three detection angles, and also at all three of the detection angles. But according to the uncertainty principle discussed in Section 3, the theory eschews commitments regarding the latter four of these probabilities:

neither the doubled pairing results

$$P\{[(A(\mathbf{a}) = +1)(B(\mathbf{b}) = +1)][(A(\mathbf{a}) = +1)(B(\mathbf{b}') = +1)]\},$$

nor

$$P\{[(A(\mathbf{a}) = +1)(B(\mathbf{b}) = +1)][(A(\mathbf{a}') = +1)(B(\mathbf{b}') = +1)]\},$$

nor

$$P\{[(A(\mathbf{a}') = +1)(B(\mathbf{b}) = +1)][(A(\mathbf{a}') = +1)(B(\mathbf{b}') = +1)]\},$$

nor the tripled pairing

$$P\{[(A(\mathbf{a}) = +1)(B(\mathbf{b}) = +1)][(A(\mathbf{a}) = +1)(B(\mathbf{b}') = +1)]$$
$$[(A(\mathbf{a}') = +1)(B(\mathbf{b}') = +1)]\}.$$

For each of these would amount to claims regarding the joint out-comes of incompatible measurements, characterised by two or three observation matrix operators $H_{\mathbf{a}*\mathbf{b}*}$ that do not commute. Quantum theory explicitly avoids such claims. That leaves four dimensions of the eight-dimensional pmf over the four detection products unspecified ... explicitly! That is why quantum theory allows four unspecified dimensions to the expectations it provides regarding the four polarisation products on the same pair of photons.

Perhaps this comment does require further explanation. You will need to view equation (12) while reading the following remarks. They concern assertions that quantum theory does allow us to make, and those that it does not. Recall that we are considering a linear programming problem in which quantum expectations are asserted for the polarisation products at the angle settings $(\mathbf{a}, \mathbf{b})$, $(\mathbf{a}, \mathbf{b}')$, and $(\mathbf{a}', \mathbf{b})$, and investigating coherent bounds for expectation of the product at the setting $(\mathbf{a}', \mathbf{b}')$. Notice firstly that quantum theory does allow us to, and indeed insists that we assert

$$E[A(\mathbf{a})B(\mathbf{b})] = q_1 + q_2 + q_3 + q_4 - q_5 - q_6 - q_7 - q_8 = 1/\sqrt{2}.$$

Examining the corresponding columns of the realm matrix seen in equation (12), it is evident that these component probabilities q_i involve assertions regarding the outcomes of both $(A(\mathbf{a})B(\mathbf{b}) = +1)$ and of $(A(\mathbf{a})B(\mathbf{b}) = -1)$, irrespective of the values of $A(\mathbf{a})B(\mathbf{b}')$ and $A(\mathbf{a}')B(\mathbf{b})$. (Look at the groupings of the first four columns of the realm matrix in (12) and of the last four columns.) Each of these two product events involve a specific outcome of the product $A(\mathbf{a})B(\mathbf{b})$ conjoined with all four possible joint outcomes of the product $A(\mathbf{a})B(\mathbf{b}')$ or over all four possible joint outcomes of the product $A(\mathbf{a}')B(\mathbf{b})$. In each case, the union of these latter

33

four outcome possibilities would constitute the "sure event". So these latter two observations incompatible with an observation of $A(\mathbf{a})B(\mathbf{b})$ would be irrelevant to the assertion of its expectation. The same feature would pertain to required assertions $E[A(\mathbf{a})B(\mathbf{b}')]$ and $E[A(\mathbf{a}')B(\mathbf{b})]$ which are also presumed in the first LP problem. None of these expectations involves any concomitant assertions regarding observations incompatible with them.

In contrast, an assertion of a probability for the *joint* occurrence of *two pairs* of polarisation product observations, such as $P\{[(A(\mathbf{a}) = +1)(B(\mathbf{b}) = +1)][(A(\mathbf{a}) = +1)(B(\mathbf{b}') = +1)]\}$ for example, would require specifications of the sum $q_1 + q_3$. Equation (12) makes clear that it is only columns 1 and 3 of the matrix in which this joint event is instantiated. Asserting a specific value for the sum $q_1 + q_3$ would necessarily entail assessments of joint probabilities for additional events that are incompatible. The same would be true of any of the other three probabilities regarding joint events for which quantum theory eschews assessment.

If one were to claim, as do the reigning proponents of Bell violations, that the probabilities of quantum theory support the valuation of $E[s(\lambda)] = 2\sqrt{2}$ according to the derivation that concluded our Section 5, that would be just plain wrong. Full stop!

Our next project is an amusing one: actually envisaging the 4-dimensional polytope of quantum probabilities relevant to the gedankenexperiment. This will be achieved by passing this 4-D quantum polytope through our 3-dimensional space, so to view it, just as the inhabitants of 2-dimensional space in *Flatland* (Abbott, 1884) viewed the sphere passing through their lower dimensional world. The sphere appeared suddenly as a point in the two dimensional plane through which it passed, and gradually expanded to circles of increasing diameter. Finally, these diminished until they suddenly disappeared again. We shall now view what we can of our 4-dimensional quantum polytope in this way.

Transforming the quantum expectation polytope into quantum probabilities

The expected photon polarisation products for which we have computed bounds can be transformed into P_{++} probabilities by applying the transformation $P_{++}(\mathbf{a}^*, \mathbf{b}^*) = [E(\mathbf{a}^*, \mathbf{b}^*) + 1]/4$ of equation (3) to the eight extreme convex hull vertices we have discovered.

This yields the vertices of another polytope in the space of the probability vector $[P_{++}(\mathbf{a}, \mathbf{b}), P_{++}(\mathbf{a}, \mathbf{b}'), P_{++}(\mathbf{a}', \mathbf{b}), P_{++}(\mathbf{a}', \mathbf{b}')]$ shown as columns of the following matrix:

$$\begin{pmatrix}
P_{++}(\mathbf{a}, \mathbf{b}) & .4268 & .4268 & .4268 & .4268 & .4268 & .4268 & 0 & .2197 \\
P_{++}(\mathbf{a}, \mathbf{b}') & .0732 & .0732 & .0732 & .0732 & .2803 & .5000 & .0732 & .0732 \\
P_{++}(\mathbf{a}', \mathbf{b}) & .4268 & .4268 & 0 & .2197 & .4268 & .4268 & .4268 & .4268 \\
P_{++}(\mathbf{a}', \mathbf{b}') & 0 & .2197 & .4268 & .4268 & .4268 & .4268 & .4268 & .4268
\end{pmatrix}.$$

The QM polytope seen passing through 3-D space

The convex hull of the 4-D column vectors shown above can be visualised through a sequence of 3-D intersections that it affords with slices perpendicular to any one of its axes. Figure 3 displays such a sequence as slices perpendicular to the $P_{++}(\mathbf{a}', \mathbf{b}')$ axis at values increasing from 0 to .4268.

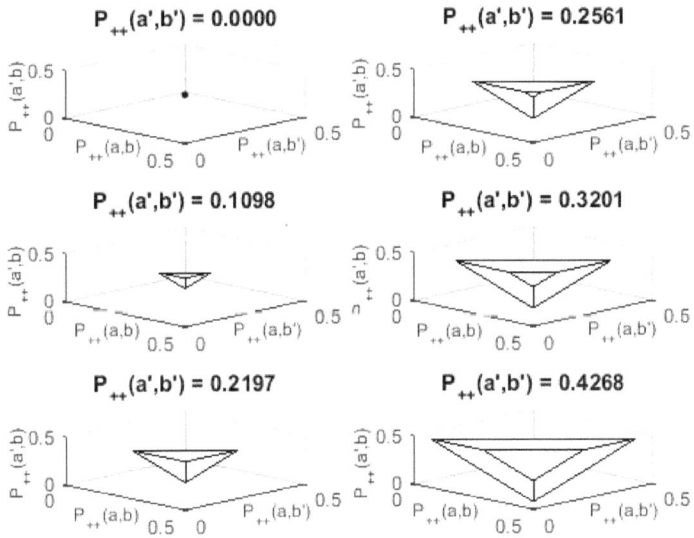

Figure 3: Sequential intersections of the 4-D convex hull of vectors $[P_{++}(\mathbf{a}, \mathbf{b}), P_{++}(\mathbf{a}, \mathbf{b}'), P_{++}(\mathbf{a}', \mathbf{b}), P_{++}(\mathbf{a}', \mathbf{b}')]$ with slices perpendicular to the $P_{++}(\mathbf{a}', \mathbf{b}')$ axis, at levels increasing from 0 to .4268.

When $P_{++}(\mathbf{a}', \mathbf{b}') = 0$, the intersection of the slice identifies only a single vertex point according to the first column of the P_{++} matrix: $[P_{++}(\mathbf{a}, \mathbf{b}), P_{++}(\mathbf{a}, \mathbf{b}'), P_{++}(\mathbf{a}', \mathbf{b})] = (.4268, .0732, .4268)$. This appears in the matrix of subplot components as the upper-left item $(1, 1)$. When the value of $P_{++}(\mathbf{a}', \mathbf{b}')$ for the slice level increases to

.1098, the intersection appears as a tetrahedron in subplot $(2, 1)$. The size of the intersecting tetrahedron increases at the $P_{++}(\mathbf{a}', \mathbf{b}')$ probability level .2197 in subplot $(3, 1)$. The tetrahedron continues to increase in size as the level of the $P_{++}(\mathbf{a}', \mathbf{b}')$ increases still further to .2561 in subplot $(1,2)$, but a corner of the intersecting polytope begins to be cut off there. This clipped portion is cut more severely from the enlarging polytope as $P_{++}(\mathbf{a}', \mathbf{b}')$ increases further, as displayed in subplots $(2, 2)$ and then $(3, 2)$, which is our view of the polytope when it suddenly disappears.

The symmetry of the configuration implies that slices along the other axes would create identical intersection sequences. (Figure 3 was produced by my colleague Rachael Tappenden who has also produced a moving sequence of this progression as the intersections proceed along the $P_{++}(\mathbf{a}', \mathbf{b}')$ axis in more refined stages.)

What to make of Aspect's empiricism

Taken in by the alluring derivation of the inequality-defying value of $E[s(\lambda)] = 2\sqrt{2}$, which mistakenly ignores the symmetric functional relations restricting the polarisation products of the gedankenexperiment, Aspect and followers were convinced that Bell's inequality is defied by quantum probabilities. The general conclusion has been that the prospect of local realism must be rejected as pertinent to quantum phenomena, supporting the view that quantum theory has identified a mysterious structure of randomness which supposedly inheres in Nature at its finest resolution. The behaviour of the photons is considered to be governed purely by a probability distribution. Aspect's programme was to devise some physical experiments that could either verify or challenge the defiance of the inequality.

According to the tenets of objective probability theory and its statistical programme, probabilities themselves are not observable quantities. What are considered to be observable are the outcomes of random variables which are generated by them. It is a matter of statistical theory to devise methods for estimating the unobservable probabilities and their implied expectations from carefully observed outcomes of the random variables they generate. Understood in this way, equation (8) constitutes a structure requiring estimation if the violation of Bell's inequality is to be verified. It is repeated atop the next page:

$$E[s(\lambda)] \quad =$$
$$E[A(\mathbf{a})B(\mathbf{b})] - E[A(\mathbf{a})B(\mathbf{b}')] + E[A(\mathbf{a}')B(\mathbf{b})] + E[A(\mathbf{a}')B(\mathbf{b}')].$$

Long-respected statistical procedures provide that the unobservable expectations of detection products on the right-hand side of this equation can be estimated by the generally applicable non-parametric method of moments. Supported by the probabilistic law of large numbers, its validity as an estimating procedure stems from the 1930s.

The programme proposed for estimating this equation proceeds by sequentially estimating the component expectations of the linear combination. To estimate the first component of $E[s(\lambda)]$, which is $E[A(\mathbf{a})B(\mathbf{b})]$, one would conduct N independent paired polarisation experiments at the angle setting $(\mathbf{a}, \mathbf{b})$, and record the value of the polarisation products $A(\mathbf{a})B(\mathbf{b})$ observed in each case, each being either -1 or $+1$. The average of these values would provide a method of moments estimate of the expectation $E[A(\mathbf{a})B(\mathbf{b})]$ which would be common to all observations in the sequence.

As described by Aspect (2002), we would conduct N repetitions of the CHSH/Bell experiment with the relative polarising angles set at $(\mathbf{a}, \mathbf{b})$, resulting in the numbers

$N_{++}(\mathbf{a}, \mathbf{b})$ observations of $(A(\mathbf{a}), B(\mathbf{b})) = (+, +)$,

$N_{+-}(\mathbf{a}, \mathbf{b})$ observations of $(+, -)$,

$N_{-+}(\mathbf{a}, \mathbf{b})$ observations of $(-, +)$, and

$N_{--}(\mathbf{a}, \mathbf{b})$ observations of $(-, -)$.

The estimated component expectation is specified as the simple average of the polarisation products in the sequence. Aspect designates this in notation formidable to a novice, as

$$\hat{E}[A(\mathbf{a})B(\mathbf{b})] \equiv \frac{[N_{++}(\mathbf{a}, \mathbf{b}) - N_{+-}(\mathbf{a}, \mathbf{b}) - N_{-+}(\mathbf{a}, \mathbf{b}) + N_{--}(\mathbf{a}, \mathbf{b})]}{[N_{++}(\mathbf{a}, \mathbf{b}) + N_{+-}(\mathbf{a}, \mathbf{b}) + N_{-+}(\mathbf{a}, \mathbf{b}) + N_{--}(\mathbf{a}, \mathbf{b})]}$$
$$(13)$$

The denominator of (13) merely equals N, the number of experiments run at this angle, displayed as the sum of its four component counts $N_{\pm\pm}(\mathbf{a}, \mathbf{b})$.

A similar programme would be followed in estimating the other three components of $E[s(\lambda)]$ pertaining to the relative angles $(\mathbf{a}', \mathbf{b})$, $(\mathbf{a}, \mathbf{b}')$, and $(\mathbf{a}', \mathbf{b}')$. An estimated version of equation (8) would then be expressed as

$$\hat{E}[s(\lambda)] \quad = \quad \text{(14)}$$

$$\hat{E}[A(\mathbf{a})B(\mathbf{b})] - \hat{E}[A(\mathbf{a})B(\mathbf{b}')] + \hat{E}[A(\mathbf{a}')B(\mathbf{b})] + \hat{E}[A(\mathbf{a}')B(\mathbf{b}')].$$

The results published in Aspect et al (1981, 1982) have been considered momentous, apparently confirming the defiance of Bell's inequality to several decimal places. What are we to make of these and subsequently embellished empirical results?

Reassessing Aspect's empirical results

Aspect reported the estimate $\hat{E}[s(\lambda)]$ from experimental data, using the method of moments as defined in equations (13) and (14). Actually, of course, it is impossible to conduct an experiment on a single pair of photons at all four angle settings, much less conduct a sequence of such experiments. Instead, experimental sequences of observations using different photon pairs were generated at each of four angle settings. These were presumed to provide independent estimates of the four expectations as they appear in the formulaic equation (13). These independent estimates were then inserted into equation (14), yielding the touted estimate $\hat{E}[s(\lambda)]$ near to $2\sqrt{2}$.

Experimentation protocols have subsequently been improved to account for the challenges of possible loopholes during the following thirty years. However, results of the estimation procedures using the improved data have been essentially identical. They have commonly been reported only in the form of so-called *p-values* of significance for hypothesis tests posed as to whether $E[s(\lambda)]$ exceeds 2 or not. The results have been celebrated as quite impressive, and deemed to be definitive.

We can now recognise the fault in Aspect's estimation procedure, which allows complete liberty in estimating the four component expected polarisation products. These use experimental incidence values of $N_{\pm\pm}(\mathbf{a}^*, \mathbf{b}^*)$ from many experimental runs *with different photon pairs at each of the angle settings*. Each of his experimental observations may be whatever value it happens to be at its angle setting, identifying whatever value of polarisation product that it does. The restrictions we have identified as instigating Bell's inequality, that each of the four products at any angle must satisfy the functional relations embedded in the results of a gedankenexperiment, have not been honoured in the process.

If Aspect's estimation procedure were meant to apply to the ontological understanding of $s(\lambda)$ in the gedankenexperiment within which he and Bell couched their theoretical claims, he would have to adjust this methodology. One might pick experimental runs using three different photon pairs at any three angles one wishes to simulate the behaviours of the products $A(\mathbf{a}^*)B(\mathbf{b}^*)$ for a single pair of photons. However, to be consistent with the Aspect/Bell problem as posed for this single pair of photons at all four relative angle settings, one then would need to compute the implied value of the polarisation product observation for the fourth angle according to the functional form that we have identified in equation (10). The same functional form would then connect the detection product at each one of the four angle settings to the other three.

Statistical estimation values reported by Aspect as well as those by subsequent research groups over the past thirty years have no relevance to the estimation of $E[s(\lambda)]$ as it is understood to pertain to four spin-products on a single pair of photons. It is perfectly reasonable to find estimation values outside of the interval $[-2, +2]$ as they have. For although these results *could reasonably pertain* to an estimate of $E[s(\lambda)]$ with $s(\lambda)$ defined as a combination of polarisation products on *four different pairs of photons*, they *do not pertain* to Bell's inequality, which is relevant to a 4-ply gedankenexperiment on the same pair of photons at all four angle settings. In the context to which their experimental results are appropriate, $E[s(\lambda)]$ is not bound by the Bell bounds of $[-2, +2]$, but rather by the interval $[-4, +4]$, which is unchallenged in this context.

Nonetheless, Aspect's empirical estimation programme might be adjusted to account for the symmetric functional relations that necessarily govern the imagined results of the gedankenexperiment. The next subsection presents the results of such an adjusted methodology. Unsurprisingly, they do not suggest any defiance of Bell's inequality at all. The simulation I construct will mimic the way Aspect's data needs to be treated, recognising his data as the result of conditionally independent experiments on distinct pairs of photons at each of the four relative angle settings of the polarisers.

Exposition by simulation

Because Aspect's experimental observation data is not available in full, I shall now expose a method for correcting his estimation procedure, and display its numerical implications using simulation data based on recognised quantum theoretic probabilities. These have found decisive empirical support both in his results at Orsay and in any number of experimental settings throughout the world.

To begin, four columns of one million (10^6) pseudorandom numbers, uniform on $[0, 1]$, were generated with a MATLAB routine. These were then transformed into simulated observations of paired photon polarisation experiments at the four relative angles we have been studying. These transformations were based on the quantum probabilities $\frac{1}{2} cos^2(\mathbf{a}^*, \mathbf{b}^*)$ and $\frac{1}{2} sin^2(\mathbf{a}^*, \mathbf{b}^*)$ we described in our equations (1). Each resulting simulated polarisation pair was then multiplied together to yield a polarisation product. In this way were created four columns of simulated observations corresponding to polarisation products from one million experiments at each of the four angles: $(\mathbf{a}', \mathbf{b}')$, $(\mathbf{a}, \mathbf{b}')$, $(\mathbf{a}, \mathbf{b})$, $(\mathbf{a}', \mathbf{b})$. They would mimic the results of the Orsay experiments. We shall refer to this matrix of simulated polarisation products below as the SIMPROD matrix.

Estimation equation (13) was then applied to each of these columns, yielding estimates of the expected polarisation product pertinent to that column, $\hat{E}[A(\mathbf{a}^*)B(\mathbf{b}^*)]$. These appear in the first row of Table 1. These four estimated expectation values were then inserted appropriately into equation (14) to yield an Aspect estimate $\hat{E}[s(\lambda)] = 2.827738$, which appears in the second row of the Table under *each* of these columns. This number is quite near to $2\sqrt{2} \approx 2.828427$, as was Aspect's reported empirical estimate, proposed as an evidential violation of Bell's inequality.

Table 1: Corrections to Aspect's estimate of $E[s(\lambda)]$

$(\mathbf{a}^*, \mathbf{b}^*)$	$(\mathbf{a}, \mathbf{b})$	$(\mathbf{a}, \mathbf{b}')$	$(\mathbf{a}', \mathbf{b})$	$(\mathbf{a}', \mathbf{b}')$
$\hat{E}[A(\mathbf{a}^*)B(\mathbf{b}^*)]$	0.707232	−0.706186	0.706840	0.707480
Aspect $\hat{E}[s]$	2.827738	2.827738	2.827738	2.827738
$\hat{E}[G(\cdot, \cdot, \cdot)]$	−0.353078	0.354348	−0.354766	−0.353934
Corrected $\hat{E}[s]$	1.767180	1.767204	1.765740	1.766964

As we now know, there is a big problem with accepting this procedure to produce an estimate of $E[s(\lambda)]$. When the product observations are supposed to apply to *the same* photon pair, the observed value of the polarisation product at any angle is required to be related to the product at the other three angles via the functional equation we exemplified our equation (10). The four of them as simulated in any row of SIMPROD *may not* all range freely in the gedankenexperiment, as they may in real experiments on *different* pairs of photons. That is a situation to which Bell's inequality does not apply. Rather, each polarisation product in the Bell setup is required to be bound to the others by the symmetric functional relation $G(\cdot, \cdot, \cdot)$ that we have identified. The rows of the matrix SIMPROD do *not* respect this requirement, so the Aspect estimate $\hat{E}[s(\lambda)]$ which they produce cannot be used to estimate the expected value of $s(\lambda)$ for the gedankenexperiment. We shall now endeavour to correct this error.

The third row of Table 1 has been generated next by producing four "Bell-restricted" columns of simulated polarisation products. This is achieved by applying the symmetric restriction function $G(\cdot, \cdot, \cdot)$ to each choice of three row components of the SIMPROD matrix. Each function value replacing only the neglected row element of SIMPROD in the row that produced it would provide a new simulated matrix of polarisation products. It is this matrix to which the Bell inequality should apply. There would be four such matrices, one for each choice of three generating columns of SIMPROD. We can call them SIMGEN matrices. The third row of Table 1 derives from the application of Aspect's estimation equation (13) solely to the restricted replacement columns so generated. The estimates in this row would constitute estimates of the expectations $E[G(\cdot, \cdot, \cdot)]$ required for a corrected estimation of $E[s(\lambda)]$ in its four representations that we have displayed as (11).

In this way we can be considered to have generated four replicas of 10^6 simulated versions of the Aspect/Bell gedankenexperiment. Their component results can be taken to be any three simulation results from a row of SIMPROD, along with the fourth result being the functionally generated result found in the same row and the appropriate fourth column of SIMGEN.

Finally, the last row of Table 1 presents "Corrected $\hat{E}[s]$" values deriving from these functional modifications of the original simulated experiments. In producing each entry from (11), the $\hat{E}[G(\cdot, \cdot, \cdot)]$ is the one appropriate to that column, while the other expected polarisation products are those appropriate to the other three columns that generated it. The elements of this row display corrected estimates of $E[s(\lambda)]$ as they should be calculated with the simulated Aspect data. Each of these four estimates is slightly different from the others. Averaging them over the four ways of generating a column of polarisation products from the other three columns of simulated products would yield a "Corrected estimate" of $E[s(\lambda)]$ as 1.766772, well within the Bell bounds of $[-2, +2]$.

Based on Aspect's report of his experimental data (Phillips and Dalibard, 2023), I feel quite sure that applying this same estimation procedure to his experimental data, considered as a simulation of the impossible gedankenexperiment, would yield a similar result.

Results on the order of this peculiar number are quite stable over repeated runs of this simulation as described. Since the theoretical analysis reported in this article yields only an interval of cohering possibilities for $E[s(\lambda)]$, this simulation leaves us with a tantalising problem of how to account for this stable result, which is quite near to $[3/\sqrt{2} - 1/(2\sqrt{2})] = 5/(2\sqrt{2}) \approx 1.767766952966369$. One might suspect that this specific result is a construct of the independence feature embedded in the simulation results across angle pairings. Such a feature might be highly suspect in Nature, given what we know now about quantum entanglement in a single experiment. However, we shall find in Chapter 5 that this same number arises in specific contexts that embed dependence relations as well. It is curious. I can also merely mention here that among all distributions in the polytope cohering with the prescriptions of quantum theory, the maximum entropy distribution inheres an expectation value of only $E(s) = 1.1522$. I have discussed this assessment and related issues recently in the *Journal of Modern Physics* (Lad, 2023), and further discussion will not detain us here. However, there can be no real empirical evidence on the issue, since it is impossible in principle to activate the setup of the four imagined simultaneous experiments on a single pair of photons. Thus, the physicists' long interest in the fabled gedankenexperiment.

Concluding comments

While Aspect's conception of statistical estimates appropriate to the photon detection problem is understandable, and corrections can be made to improve its relevance to the Aspect/Bell problem, developments of statistical theory and practice during the past fifty years have surely generated superior methods for evaluating the physical theory of quantum behaviour. These rely on the subjective theory of probability which has gained substantial credibility from the past half-century of research in the foundations of probability and statistics, under the leadership of Bruno de Finetti and researchers adhering to his viewpoint. There are even some prominent physicists among its proponents, though not many.

Proclaimers of inherent randomness in the physics of quantum behaviour have long reigned in the forum of scientific opinion, largely on the basis of the mistaken violation of Bell's inequality that we have now debunked.

The mathematical structure of the Aspect/Bell problem and its resolution align well with the theory of subjective probability. This viewpoint is in keeping with Einstein's interpretation of quantum mechanics, known by his famous adage that the old one does not roll dice. An expansion of relevant ideas in the context of the constructive mathematics of Bruno de Finetti's operational subjective statistical method can be found in an article of Romano Scozzafava (2000). It is true that readers more comfortable with the standard realist interpretation of quantum mechanics may still consider the probabilities as ontic properties of the photons themselves, without disturbing the mathematical issues we have engaged. However, such a view can be favoured only on the basis of pure belief. There is no physical experiment that can resolve the claims inhering in the setup of the gedankenexperiment underlying Bell's inequality. Of one thing we can now be sure: the theory of quantum mechanics does not support its defiance.

As to the characterisation of the theory of hidden variables, this is another endeavour that has been misconstrued in accepted literature, largely on the basis of the mistaken understanding of the defiance of Bell's inequality that we have corrected here. We shall examine this matter technically in Chapter 3, finding that the proposition of supplementary variables might apply to any distribution of uncertainty over observations of polarisation experiments whatsoever. No coherent distribution over observable quantities supports the defiance of Bell's inequality, whether considered to be a formalisation of hidden variables theory or not.

Virtually all discussion of quantum probabilities since the original work of Bell has supported the conclusion that probabilities pertinent to quantum behaviour can violate the seemingly innocuous inequality that he identified. The mathematical error that has been discovered and reported here substantiates the end of an era of accepting this conclusion. The results we have aired will have ramifications for many published estimations based on more sophisticated experimentation as well. Further consequences for a host of theoretical issues have been studied and discussed in the context of the mistaken understanding. These include the related notions of entangled particles, information transfer, and even the many-worlds hypothesis. Discussion of these topics do require philosophical attention to a variety of conceptual constructs in which they are embedded. However, the analysis of Aspect/Bell presented here has nothing to do with philosophical distinctions. It has identified a mathematical error in accepted work that must be recognised no matter what might be the philosophical positions of interested parties. Probabilistic forecasts motivated by quantum theory do not violate any laws of probability theory. Full stop.

There is a substantive literature of technical challenge to the mainstream understanding that I have not attempted to review in this introductory chapter, including an important article of Adenier (2001). This recognises quite clearly the relevance of Bell's inequality exclusively to a gedankenexperiment on the same pair of photons at all four relative polarisation angles, something which many deny, despite its insistence by Aspect. Nonetheless, the results we have assessed herein are truly novel: an identification of the error in the accepted computation of $E[s(\lambda)]$ as $2\sqrt{2}$; recognition of the symmetric functional relations among the four polarisation products involved in the CHSH form of the problem; exposition of the 4-D polytope of cohering quantum theoretic distributions for the components of the 4-ply quantity s; and identification of the quantum theoretic interval estimate of $E(s)$ as $(1.1213, 2]$.

Perhaps surprisingly, Aspect/Bell's was just the beginning of a sequence of egregious errors by quantum theorists. We shall elucidate them in the remainder of this book, beginning with a stunner in the next chapter.

Chapter 2

EXTENDING THE SETUP TO
TWO PAIRS OF PARTICLES:
a gedankenexperiment
requiring more denken

When I first began to disseminate the results of my analysis that refute the supposed defiance of Bell's inequality by quantum probabilities, I was immediately challenged by suspicious and offended physicists. Dismissed as incomprehensible and confused, the focus of my analysis was regarded as completely out of date with subsequent developments. Not only was Aspect's experimental setup deemed to be crude and to have been superseded, but it was proposed to me that a definitive analysis by Greenberger, Horne, Shimony and Zeilinger (1990) (hereafter GHSZ) had determined that the premises of the EPR argument entailing "perfect correlation, reality, locality, and completeness" could not even be modelled consistently in quantum problems involving more than two dimensions. GHSZ were said to have expounded a version of Bell's theorem without involving inequalities. Aware of my evident naivety regarding matters I had long avoided in my researches, and keen to learn what this might be about, I eagerly dove right in!

Upon reading, I was stunned to learn that their claims, already studied and honoured within the physics community for some thirty years, are based on an argument that defies a simple rule of deductive logic, and are entrained in cryptic notation that has misled the leading researchers of the scientific community. Their article will have been studied by surely more than five thousand physicists and students during this time. It is with some embarrassment that I shall explain these miscues in this chapter. I invite you to please continue reading, with an open mind, if only out of curiosity.

Any reader prepared to do so is invited to become acquainted (or re-acquainted) with the GHSZ argument in their own words, before continuing. However, to make this presentation self-contained, their argument will be developed here faithfully until the point that it hits the fan, using their notation precisely until it will become necessary to refine and embellish it. I use their numbering system for equations, beginning with (7), the first numbered equation of their Section III, which is the sole subject matter of this exposition. When their equations are repeated using a refined notation, their same equation numbers are merely adorned with a superscripted star *. Lower case Roman numerals are employed to designate the few new equations that I introduce. I mention in passing only that the persuasive exposition of Bell's original Theorem in the Section II of the GHSZ article continues to embed the error of neglect we have exposed in Chapter 1, which Bell and subsequent advocates of the supposed violation had missed.

My deferential attitude to your invited review of the original paper, and the care with which I shall now assess it, are appropriate to the influential role the article has played in motivating the current general misunderstanding of the quantum scale defiance of Bell's inequality within the physics community. Of course the inequality itself has provoked extensive analysis and a myriad of commentaries. Notable among them are the comprehensive review by Brunner et al (2014), and wide-ranging discussions in the texts of Greenstein and Zajonc (2006) and of Jaeger (2009). Within this context, the provocative claims of GHSZ to a contradiction involved in the formulation of theories of local realism are considered to be definitive. They form part of a stimulating literature that stems from the initial challenging work of von Neumann (1932, 1955) and runs through Mermin and Schack (2018). While subscription to various arguments has varied over time, the concluding assessment of GHSZ by Jaeger (2009, p. 44-45), reproduced below, is supported by virtually all researchers who have reviewed it:

> Remarkably, a decade after Aspect's tests of the CHSH inequality, it was shown by Greenberger, Horne, Shimony, and Zeilinger (GHSZ) that the premises of the

Einstein-Podolsky-Rosen paper become inconsistent when applied to systems possessing three or more subsystems, even for the cases involving such perfect correlations [194]. The GHSZ demonstration shows that the incompatibility of the EPR assumptions with quantum mechanics is stronger than that indicated by the violation of the Bell and CHSH inequalities, in that in the case of a pair of two-level systems there is no internal contradiction at the level of perfect correlations. Indeed, Bell produced an explicit model for the case of a pair of spin-1/2 particles demonstrating the consistency of the EPR conditions with the perfect correlations predicted by quantum mechanics [23]. Furthermore, the contradiction between quantum mechanical predictions and the Bell and CHSH inequalities are expressions violated only by statistical predictions of quantum mechanics, rather than by individual events. In the lead up to the exceptionally clear exposition of GHSZ, Greenberger, Horne, and Zeilinger (GHZ) demonstrated the inconsistency in a new way in systems consisting of three or more correlated spin-1/2 particles [195]. Because this showed that the incompatibility of quantum mechanics with the EPR assumptions arises at the level of perfect correlations rather than statistical predictions and did not require the use of an inequality, these results are often referred to as "Bell's theorem without inequalities."

In a word, I am fully aware of the challenging content of the analysis I shall present here, and the extent of the reconsideration it should require. While my presentation will be lively, it is not with bravado but with honest concern that I suggest we have been deeply engaged in a serious mistake for some fifty years since the advent of Bell's first publications on the matter. This was a mistake he himself had suspected but could not identify. If I am in error, it will need to be displayed by reasoning rather than by reverence for vaunted old understandings. We all make mistakes. Let's get into it.

Physical setup of the GHSZ experiment

The context of the GHSZ experiment does not involve the behaviour of photons, but rather of four electrons that are propelled toward Stern-Gerlach analysing magnets so to determine the directions of their orbital spins. The physical setup is caricatured by Figure 2 of GHSZ, which is reproduced below.

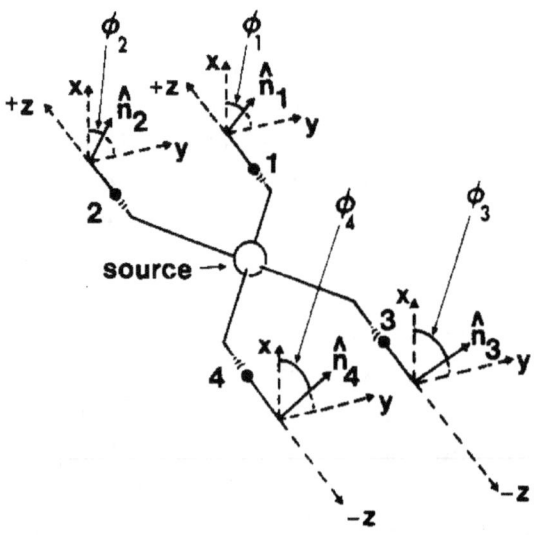

Fig. 2. A four-particle gedankenexperiment. The source emits a quadruple of spin-1/2 particles, 1,2,3, and 4, in the state of Eq. (7). Particle i ($i = 1,2,3,4$) enters its own Stern–Gerlach apparatus oriented along direction $\hat{n}_i$. We emphasize that the four Stern–Gerlach apparatuses can be separated by arbitrarily large distances. Behind each apparatus two detectors, not shown, record whether the result is up or down.

Reproduced from *American Journal of Physics*, **58**(12), p 1134, doi.org/10.1119/1.16243, with the permission of the American Association of Physics Teachers.

This schema represents a physical process in which one pair of electrons (referred to as "spin-1/2 particles") are transmitted in spatially separated beams in the direction $^{+}$z toward Stern-Gerlach analysers at stations 1 and 2 while another pair are transmitted far away in the opposite direction $-$z toward similarly separated analysers at stations 3 and 4. The electrons are said to be in a superposition of two states vis-à-vis their spins, implying that they might be observed as either "up" or "down" when detected at any analyser. As transmitted, they are considered to be in both and neither states, superposed. The observation "up" is designated by a statistic recorded as $+1$, while "down" is recorded as -1.

The Stern-Gerlach magnets can be positioned in various directions relative to that in which the incoming particles arrive. These directional vectors are denoted in Figure 2 as $\hat{n}_1, \hat{n}_2, \hat{n}_3$, and $\hat{n}_4$. In the (x, y) plane perpendicular to any z direction, the Figure denotes by ϕ the angle relative to the $x-$dimension in which any one of them is directed. The four angle sizes ϕ_1, ϕ_2, ϕ_3, and ϕ_4 can be variable, but they are fixed at specific values for any run of the experiment. Of course, the (x, y) planes of these detectors could also be tilted toward or away from the direction of their z axes as well. Such can be accounted for in the theory according to an equation, numbered as (8) in the GHSZ article, in which these angles are denoted by θ_i. However, they are not depicted in the Figure, for throughout this discussion they are considered to be fixed at $\theta_i = \pi/2 = 90°$, each of the four (x, y) planes being perpendicular to the direction of z. This reduces their general equation (8) to their (9) which we shall examine shortly.

The 4-component electron propagation mechanism, which we shall not discuss here, ensures the spin state of the four-plex of electrons is superposed according to the prescription numbered (7):

$$|\Psi\rangle \ = \ (1/\sqrt{2}) \ [\ |+\rangle_1 \ |+\rangle_2 \ |-\rangle_3 \ |-\rangle_4 \ + \ |-\rangle_1 |-\rangle_2 |+\rangle_3 |+\rangle_4 \ ] \ .$$

If you are not au fait with the mathematical apparatus of quantum theory, please do not be put off by this notation, and read on. This is merely a concise notation for representing the structure of the experimental situation we have explained in the previous paragraph. Trust yourself. You are ready to go.

Identifying the prescriptions of quantum theory

When detections are made of the electrons' spins at stations 1, 2, 3, and 4, these are designated by the variable names A, B, C, and D, respectively. Each of these values might arise as -1 or $+1$ according to the direction of the corresponding spin observation. Now what would be the values of these observation values in any run of the experiment? The theory of quantum mechanics does not specify exactly what the results would be in any run of such experiments, even though each run in a sequence might be set up carefully in exactly the same way. Quantum theory yields only a probability distribution for the four measurement results from any run. Just as our investigation of the polarised photon pair detection in the Aspect/Bell experiment found the crucial feature of the observations was the value of the *product* of the detection results, the crucial part of the quantum theoretic specifications of probabilities for observation values of the four electron spins in a run revolves about their product, now a product of four: $ABCD$.

The product of the spin measurements in any run depends stochastically on the four angle settings in the design of that run. GHSZ derive the mathematical expectation of this spin-product in Appendix F of their article for the general case in which the rotational angles θ_i of the magnets toward the incoming electrons are variable. In the article itself they simplify this expectation to the special case in which each of the (x, y) planes is perpendicular to its relevant z axis. Using cryptic notation, the spin-product expectation is reported for this setup in a numbered equation as

$$E^{\psi}(\hat{\mathbf{n}}_1, \hat{\mathbf{n}}_2, \hat{\mathbf{n}}_3, \hat{\mathbf{n}}_4) \;=\; -cos(\phi_1 + \phi_2 - \phi_3 - \phi_4) \ . \tag{9}$$

They describe this equation as "the expectation of the product of the outcomes when the orientations are as indicated".

To be sure, this notation has a sensible motivation. For it pertains to an expectation of a function (the product $ABCD$) of spin observations on a system of particles. These are designed to reside in the superposed state designated by $|\mathbf{\Psi}\rangle$, in which the Stern-Gerlach analysers are positioned at angles corresponding to the directional vectors $\hat{\mathbf{n}}_1, \hat{\mathbf{n}}_2, \hat{\mathbf{n}}_3$, and $\hat{\mathbf{n}}_4$. These directional vectors determine the angles ϕ_1, ϕ_2, ϕ_3, and ϕ_4 whose linear combination is the argument of the cosine function in (9) that specifies the expectation value. Well, I do not think I am being picayune in highlighting here that this expectation $E^{\psi}(\hat{\mathbf{n}}_1, \hat{\mathbf{n}}_2, \hat{\mathbf{n}}_3, \hat{\mathbf{n}}_4)$ is *not* the expectation of a vector of four directional vectors (which it ostensibly is), but rather it is the expectation of the product of four spin observation values in an experiment at which the analyzing magnets are aligned with these directions. It would be depicted more clearly using the designation $E^{\psi}[ABCD(\phi_1, \phi_2, \phi_3, \phi_4)]$, which would be standard in mathematical probability and statistics. While understandable, the notation that GHSZ use serves to obscure a sensible analysis of the situation, as we shall now discover.

Of particular interest to the GHSZ argument are the cases of perfect correlation, which they designate by the conditioned equations

$$\text{If} \quad \phi_1 + \phi_2 - \phi_3 - \phi_4 = 0$$

$$\text{then} \quad E^\psi(\hat{\mathbf{n}}_1, \hat{\mathbf{n}}_2, \hat{\mathbf{n}}_3, \hat{\mathbf{n}}_4) = -1\,, \tag{10a}$$

$$\text{If} \quad \phi_1 + \phi_2 - \phi_3 - \phi_4 = \pi$$

$$\text{then} \quad E^\psi(\hat{\mathbf{n}}_1, \hat{\mathbf{n}}_2, \hat{\mathbf{n}}_3, \hat{\mathbf{n}}_4) = +1\,. \tag{10b}$$

The conditioned equations are printed exactly like that, in four lines of which the second and fourth lines conclude with the equation number identifications $(10a)$ and $(10b)$. This unfortunate feature of the double column print style of their publication will be found relevant to their mistaken argument. The numbered equations do not stand on their own, but each relies upon the conditioning qualifier, which precedes it on an unnumbered separate line! The two conditions are contradictory. Either one of them or neither of them may hold in any specific experimental run, but not *both* of them.

A second unfortunate feature of their considerations is that they did not designate explicitly in notation the statement that they described in words: that each multiplicand of the single observed product $ABCD$ of the spin recordings depends (stochastically) on the *four* orientations of the detection magnets, $(\phi_1, \phi_2, \phi_3, \phi_4)$. Mentioning for now only that these two misfortunes will return to haunt us, we shall continue with the GHSZ argument.

Interest in the conditions of $(10a)$ and $(10b)$ arises from the fact that they support the supposition of Einstein, Podolsky, and Rosen (EPR, 1935) of perfect correlation, a condition specified by the derivations of quantum theory. The other three suppositions of EPR (reality, locality, and completeness) are noted to concern matters extraneous to quantum theory, though they are recognised as plausible and in agreement with principles of classical physics.

GHSZ were pleased that this experimental design might instantiate the requirements of Bell's inequality in a four-dimensional problem. However, they proceeded no further in pursuing details of this matter, because they first pursued an argument that the structure they had developed to this point embeds a contradiction among the four premises of EPR. These were the premises of (i) perfect correlation, (ii) reality, (iii) locality, and (iv) completeness. Let's begin to follow their line of argument.

Pursuing the contradiction "discovered" by GHSZ

They begin by restating the conditional statements $(10a)$ and $(10b)$, but now using functional observation notations $A(\cdot)$, $B(\cdot)$, $C(\cdot)$, and $D(\cdot)$ in stating the conclusions of the "If" clauses. They write

$$\text{If} \quad \phi_1 + \phi_2 - \phi_3 - \phi_4 = 0$$
$$\text{then} \quad A_\lambda(\phi_1)B_\lambda(\phi_2)C_\lambda(\phi_3)D_\lambda(\phi_4) = -1, \quad (11a)$$
$$\text{If} \quad \phi_1 + \phi_2 - \phi_3 - \phi_4 = \pi$$
$$\text{then} \quad A_\lambda(\phi_1)B_\lambda(\phi_2)C_\lambda(\phi_3)D_\lambda(\phi_4) = +1. \quad (11b)$$

Before continuing with their development, I must make two remarks, either of which might be ignored, but both of which are of meritorious consequence.

The first concerns an unspoken argument. Notice that the lines numbered $(10a)$ and $(10b)$ make statements about the values of *expectations of the products* of the spin measurements at the four observation sites, whereas lines $(11a)$ and $(11b)$ concern the *products of observations themselves*. Upon consideration, this appears to be of little consequence. Why? The numerical values of each multiplicand can be only either -1 or $+1$ by their very operational definition, so the product of any four such observations may also equal only either -1 or $+1$. Moreover, equation $(10a)$, for example, quite rightly designates that under the condition specified in the unnumbered line just above it, the expectation must equal -1. This implies that the probability weight attributed to the possibility that the product equals $+1$ must be zero! Under this condition, the product of the spins itself must equal -1. Thus, quantum theory prescribes that in this case, the *product* of the four spin observations itself must equal -1, just as stated in equation $(11a)$. The very same remark would pertain to the replacement of $(10b)$ by $(11b)$, according to which the product of the four observations must equal $+1$.

Thus, changing the statement of equations $(10a, b)$ regarding expectations of a function (the product of four measurements) to $(11a, b)$ regarding the function values themselves appears innocuous here, because the distributions of the function value are obviously degenerate in both cases, at -1 and $+1$ respectively. However, we should be aware that the joint distributions of the *component multiplicands* of this product have *not* been determined to be degenerate. There are several arrays of the multiplicands that could yield a product of -1 to instantiate $(11a)$, and several arrays that could yield a product of $+1$ so to instantiate $(11b)$. It is the expectation of their *product* that equals -1, and thus it is their product that has a degenerate distribution. The distribution of the four component observations in a run remains ample. Enough said for now, but we shall return to this recognition later.

The second remark is less consequential, but it is substantive relative to the context of the GHSZ analysis. At this point in their development, the spin observation variables A, B, C, and D suddenly begin to appear with the subscript λ, as in $A_\lambda(\phi)$. As it turns out, *every* subsequent appearance of such observation values in their article is adorned with this subscript, without further comment. Since the subscript contributes nothing to the force of the relevant arguments, I will no longer use it. To be sure, its presence would serve as a reminder that what we are doing here is formalising experimental procedures motivated by considerations of the EPR proposal of "supplementary variables" as the source of the incompleteness of quantum theory. The vector λ is meant to designate the unknown values of such variables. Now this *is* a very important matter bearing consideration, and we shall investigate this matter explicitly on its own in the next chapter. For now, however, let us consider ourselves to have been reminded of this important motivation, and just drop the λ subscript, except when I quote GHSZ verbatim. For there will come a point in these considerations at which we shall wish to use the subscript position on the spin measurements to make explicit another important *varying* feature of the problem that has been ignored hitherto.

Continuing then from equations $(11a)$ and $(11b)$, I shall follow very closely the work and even the wording of GHSZ. Exact quotations will be marked.

Next are considered some implications of $(11a)$, a numbered equation that holds under the condition that the combination of magnet angles $\phi_1 + \phi_2 - \phi_3 - \phi_4 = 0$. Four specific instances of this specification are proposed, with reference to any arbitrary angle size ϕ, as

$$A(0)B(0)C(0)D(0) = -1 \qquad (12a)$$

$$A(\phi)B(0)C(\phi)D(0) = -1 \qquad (12b)$$

$$A(\phi)B(0)C(0)D(\phi) = -1 \qquad (12c)$$

$$A(2\phi)B(0)C(\phi)D(\phi) = -1, \qquad (12d)$$

because the four angle arguments in each of these product equations meet the condition that $(\phi_1+\phi_2-\phi_3-\phi_4) = 0$. This is the condition under which $(11a)$ holds.

As a consequence of equalities $(12a)$ and $(12b)$ they then obtain

$$A(\phi)C(\phi) = A(0)C(0) \qquad (13a)$$

through cancellation of $B(0)D(0)$ in the equations' quotient. This appears identically on the left-hand sides of these two equations. Correspondingly, equalities $(12a)$ and $(12c)$ are seen to imply

$$A(\phi)D(\phi) = A(0)D(0) \qquad (13b)$$

according to similar cancellations of $B(0)C(0)$.

Then, the quotients of corresponding sides equations $(13a)$ and $(13b)$ yields them

$$C(\phi)/D(\phi) = C(0)/D(0) , \qquad (14a)$$

which can be rewritten as

$$C(\phi)\, D(\phi) = C(0)\, D(0) , \qquad (14b)$$

because both $D(\phi)$ and $D(0)$ can each equal only either $+1$ or -1. In either case, each of them is equal to its arithmetic inverse.

By substituting $(14b)$ into $(12d)$, GHSZ then obtain

$$A(2\phi)B(0)C(0)D(0) = -1 . \qquad (15)$$

Combining this in a quotient with $(12a)$, which says that $A(0)B(0)C(0)D(0) = -1$, yields them the surprising result that

$$A(2\phi) = A(0) = const \text{ for any angle } \phi. \qquad (16)$$

In particular, this would imply that a measurement of $A(\pi)$ must equal $A(0)$. (Just set ϕ equal to $\pi/2$.)

While this result is apparently not contradictory itself, to a physicist well versed in quantum theory it is surely quite troublesome. GHSZ call it a "preliminary result". Let them explain why it appears troublesome in their own words. "For if $A_\lambda(\phi)$ is intended, as EPR's program suggests, to represent an intrinsic spin quantity, then $A_\lambda(0)$ and $A_\lambda(\pi)$ would be expected to have opposite signs." [Electron spins measured by analysers positioned in two opposing directions, differing in orientation by π radians $= 180°$, should display opposite directions rather than the same direction, just as two attracting magnets will repel one another if one of them is rotated by $180°$.] However, this conundrum is subjected to no further examination, on account of a really stunning argument.

Witness the sleight of hand!

GHSZ continue. "The trouble becomes manifest, and an actual contradiction emerges, when we use $(11b)$ — which until now has not been brought into play — to obtain

$$A_\lambda(\theta + \pi)B_\lambda(0)C_\lambda(\theta)D_\lambda(0) = 1 \qquad (17)$$

which in combination with Eq. $(12b)$ yields

$$A_\lambda(\theta + \pi) = -A_\lambda(\theta). \qquad (18)$$

"This result *confirms* the sign change that we anticipated on physical grounds in EPR's program, but it also *contradicts* the earlier result of Eq. (16) (let $\phi = \pi/2, \theta = 0$). We have thus brought to the surface an inconsistency hidden in premises (i)-(iv)."

N.B. Their parenthetical suggestion means to let $\theta = 0$ in (18) and to let $\phi = \pi/2$ in (16).

What could be a more stunning demolition of the EPR premises?

Answer: some calm logical thinking, and the truth!

Well, what could be wrong here?

A little more denken

To begin with a compelling observation, neither of the equations numbered $(11a)$ and $(11b)$ stands on its own. Each of them is a conclusion of a conditioning clause, an "if clause"; and these two clauses are quite evidently contradictory. Yet GHSZ use these two conclusions in concert. Equation $(11a)$ results from quantum theory applied to a situation in which the four magnet angles are designed to meet the condition "If $\phi_1 + \phi_2 - \phi_3 - \phi_4 = 0$". Equation $(11b)$ provides the conclusion to the condition "If $\phi_1 + \phi_2 - \phi_3 - \phi_4 = \pi$". If either of these conditions holds for an experiment under consideration then the other *cannot*. The two conditions cannot be instantiated at the same time, and they may not be honoured at the same time ... and neither may the the distinct conclusions they motivate. It is surely not permitted to combine their equation (17) with equation $(12b)$ to yield (18). Professors Greenberger, Horne, Shimony and Zeilinger discover the contradiction in the EPS suppositions that they do *because they have introduced the contradiction into their analysis themselves!* Full stop.

A brief interlude

Compounding the misunderstanding this analysis has supported, empirical results from an actual experiment in a companion context of three photon polarisation observations was presented by Pan, Bouwmeester, Daniell, and Zeilinger (2000). It claimed to prove the inadequacy of a locally realistic model to account for them. Subsequently, this was challenged by Aschwenden et al. (2006), who proposed just such a model that improved the experimental explanation provided by a purely quantum theoretic model. However, the problems with the empirical programme engaged by Pan et al. run even deeper than the empirical challenge of the Aschwenden group. I mention these works here because the claims of Pan's group must be recognised and addressed, given their influential importance in the literature of quantum physics. However, a full analysis would cover ground with details not appropriate to this book. To avoid a distraction here, I defer a terse commentary to an Appendix at the end of this chapter. My comments there are meant only for specialists who are already deeply imbued in the argument they have presented. You may examine it once you finish this chapter, or ignore it, as befits your interest and expertise. We'll continue.

Return to the trail

Returning to our assessment of the GHSZ article itself, any serious reader should have been startled to observe the *joint* supposition of the contradictory premises (11a) and (11b) when the authors follow the implications of both equations together to yield the contradiction they claim to have "found". So it is astounding to recognise the acclaim that has been afforded by physicists to this spurious result. However, the confusions in the GHSZ argument run deeper still. There is much more that can be learned by continuing a pursuit of the "troublesome preliminary result" (16) from which their supposed discovery of a contradiction then deterred them. We shall learn now that the source of this result derives from another serious error. Less evident to a new reader, perhaps, it is nonetheless shocking, having arisen from insufficient thinking combined with the use of casual notation. Let's look into it.

Pursuing the trouble with complete notation

I had initially been taken in by the GHSZ argument, and I shared their puzzlement. However, whereas they were concerned with the identical signs of spins observed from two opposing directions, I wondered how might the angle ϕ_1 at station A equal both 2ϕ *and* 0 so as to instantiate equation (16) for any angle value ϕ? The quantities $A(2\phi)$ and $A(0)$ explicitly denote spin observations at station 1 in two different experiments in which the orientation of the detection magnet at station A differs. The outcome of the experiment is random in either of them, assessed with quantum probabilities that depend on the magnet angles at the other three stations as well, but allowing each of them to equal either -1 or $+1$. There is no requirement of any sort that these outcomes need be identical in the two *different* experimental runs. What could their equation $A(\pi) = A(0)$ mean? It turns out that there is a clear way out of either conundrum, as we shall see. By reconsidering the development of their equation (16), we can find how to clarify the situation. Sad to say it, we shall need to start at the beginning.

The GHSZ argument was proposed some fifty years after the recognition of so-called particle entanglement had arisen among quantum theorists. This had made all of us aware that consideration of any aspect of particle behaviour occurring at station 1 with an observation value labelled A will typically depend on both the settings of the angles *at all four stations*, and depend on the behaviours of the particles at the other three stations as well. I mention this because at the very start of their argument, GHSZ casually exhibit two particular instantiations of their expectation equations $(10a, b)$ by writing

$$\text{If} \quad \phi_1 + \phi_2 - \phi_3 - \phi_4 = 0$$
$$\text{then} \quad A_\lambda(\phi_1)B_\lambda(\phi_2)C_\lambda(\phi_3)D_\lambda(\phi_4) = -1, \qquad (11a)$$
$$\text{If} \quad \phi_1 + \phi_2 - \phi_3 - \phi_4 = \pi$$
$$\text{then} \quad A_\lambda(\phi_1)B_\lambda(\phi_2)C_\lambda(\phi_3)D_\lambda(\phi_4) = +1. \qquad (11b)$$

Equations $(10a, b)$ had quite rightly designated the expectations in their concluding clauses (the left-hand sides of the equations in their "then" statements) as functions of four directional variables, designated by $E^\psi(\hat{\mathbf{n}}_1, \hat{\mathbf{n}}_2, \hat{\mathbf{n}}_3, \hat{\mathbf{n}}_4)$. Although I have already remarked about the cryptic form of this notation, their verbal description did have the feature of recognising them as expectations of a general function of four variables, the directional vectors of the four magnet settings. However, in their instantiation equations $(11a, b)$ GHSZ blithely represent this product function $ABCD(\phi_1, \phi_2, \phi_3, \phi_4)$ as a separable function of four independent variables standing alone in singular directional settings: $A(\phi_1)B(\phi_2)C(\phi_3)D(\phi_4)$. For the moment I say only, "Beware!" But this misconstrual of the situation becomes even more abusive, and we shall need to look into it deeply to uncover the havoc it has promoted.

We need to designate the arguments of the spin measurement functions A, B, C, and D completely if we are to denote fully the experimental context in which the observations are made. The measurement which GHSZ denote simply as $A_\lambda(\phi_1)$ requires embellishment to $A_\lambda(\phi_1, \phi_2, \phi_3, \phi_4)$ if it is to denote an observation of the spin A on the first of four entangled particles. The magnet angle at station 1 may well have been set at ϕ_1, but what were the directional angles for the spin detectors of the other three particles with which the particle entering station 1 is entangled? This surely makes a difference, as the distinction between equations $(11a)$ and $(11b)$ makes clear. On the one hand, one might might insist on the full experimental design notation $A(\phi_1, \phi_2, \phi_3, \phi_4)$ in the identification of the spin measurement A in any experimental run. However this would amount to an ungainly designation of a spin-product of four such measurements. I am going to suggest and to follow henceforth a notation that will use the subscript position under A to designate the full angular context in which a measurement of A at its station angle ϕ_1 is made. So I shall write $A_{(\phi_1, \phi_2, \phi_3, \phi_4)}(\phi_1)$ to designate this. Remember that we are forgoing the unvarying GHSZ universal subscript of λ on spin values A, B, C, and D, except at times when I quote them exactly.

Using this notational convention, we need to rewrite the GHSZ functional form equations $(11a, b)$ of the general expectation result $(10a, b)$, in the form

$$\text{If} \quad \phi_1 + \phi_2 - \phi_3 - \phi_4 = 0$$

$$\text{then} \quad A_{(\phi_1, \phi_2, \phi_3, \phi_4)}(\phi_1) B_{(\phi_1, \phi_2, \phi_3, \phi_4)}(\phi_2)$$
$$C_{(\phi_1, \phi_2, \phi_3, \phi_4)}(\phi_3) D_{(\phi_1, \phi_2, \phi_3, \phi_4)}(\phi_4) = -1, \quad (11a)$$

$$\text{and} \quad \text{If} \quad \phi_1 + \phi_2 - \phi_3 - \phi_4 = \pi$$

$$\text{then} \quad A_{(\phi_1, \phi_2, \phi_3, \phi_4)}(\phi_1) B_{(\phi_1, \phi_2, \phi_3, \phi_4)}(\phi_2)$$
$$C_{(\phi_1, \phi_2, \phi_3, \phi_4)}(\phi_3) D_{(\phi_1, \phi_2, \phi_3, \phi_4)}(\phi_4) = +1. \quad (11b)$$

My apologies for this gaudy notation, but we do require it to air an egregious error in the GHSZ argument, to which we now turn.

Appearing now *somewhat* less ungainly, the specific instantiation equations they enumerate as equations $(12a, b, c, d)$ would be designated in this form as

$$A_{(0,0,0,0)}(0)B_{(0,0,0,0)}(0)C_{(0,0,0,0)}(0)D_{(0,0,0,0)}(0) \;=\; -1 \quad (12^*a)$$

$$A_{(\phi,0,\phi,0)}(\phi)B_{(\phi,0,\phi,0)}(0)C_{(\phi,0,\phi,0)}(\phi)D_{(\phi,0,\phi,0)}(0) \;=\; -1 \quad (12^*b)$$

$$A_{(\phi,0,0,\phi)}(\phi)B_{(\phi,0,0,\phi)}(0)C_{(\phi,0,0,\phi)}(0)D_{(\phi,0,0,\phi)}(\phi) \;=\; -1 \quad (12^*c)$$

$$A_{(2\phi,0,\phi,\phi)}(2\phi)B_{(2\phi,0,\phi,\phi)}(0)C_{(2\phi,0,\phi,\phi)}(\phi)D_{(2\phi,0,\phi,\phi)}(\phi) \;=\; -1 \,. \quad (12^*d)$$

The station angle vectors that subscript the multiplicands in each one of these products meet the condition that $\phi_1 + \phi_2 - \phi_3 - \phi_4 = 0$, the condition under which $(11a)$ holds.

As an appeasement to your understandable hopes for a simpler notation, I can offer that there will be some contexts in which it will be sufficient to subscript a spin observation with only the value of $\kappa \equiv \phi_1 + \phi_2 - \phi_3 - \phi_4$, writing $A_{\kappa=0}(0)$ in place of $A_{(0,0,0,0)}(0)$. However in the four lines of (12^*), the experimental contexts are such that every spin observation would then be subscripted with $\kappa = 0$. This would mean, for examples, that $D_{(0,0,0,0)}(0)$ in (12^*a) and $D_{(\phi,0,\phi,0)}(0)$ in (12^*b), would both be designated by $D_{\kappa=0}(0)$, whereas in fact they quite rightly designate very different things, being the observation values of D in two quite different experiments and experimental settings. There is no assurance at all that they will instantiate at the same numerical value, without presuming local realism. This is a constraining condition that we shall formalise soon.

GHSZ casually ignore this whole contextual matter about which their knowledge of quantum entanglement should have alerted them. They denote both $D_{(0,0,0,0)}(0)$ and $D_{(\phi,0,\phi,0)}(0)$ by $D_\lambda(0)$, and presume that they are always equal in any runs of their experiment, feeling free to cancel them against one another when they do their algebra. Then they are briefly surprised by their equation (16) which results. However, as they carry on with their self-contradictory analysis, they ignore the entire issue, surprising preliminary result and all. To the contrary, we shall now trudge into these matters in great gory detail. However, we need to first provide a clarifying assessment of the entire setup.

Clarification via examination of realm matrices

For clarification of a central aspect of the situation relevant to all that follows, equation (i) below displays the ensemble of observation possibilities (which I call a "realm matrix") for an observable spin vector arising in the conduct of a specific Stern-Gerlach experiment on four entangled particles. The observations might be made on four particles in a quantum state $|\Psi\rangle$ corresponding to *any* specific experimental design for which $\phi_1 + \phi_2 - \phi_3 - \phi_4 = 0$. To repeat, this condition is a supposition ("If clause") of GHSZ equation $(11a)$, which does not stand on its own without this clause. Notice that the companion equation $(11b)$ relies on an *alternative* condition, that $\phi_1 + \phi_2 - \phi_3 - \phi_4 = \pi$. Contradictory to each other, both of these conditions cannot be satisfied in any specific experimental run, nor even in the imagined runs of a gedankenexperiment on a specific single quartet of particles. The angle combination may equal either 0 or π in any exerimental setup, but it cannot equal both at the same time. Nor can their two concluding equations hold at the same time, if we are to insist on discussion that honours the prescriptions of deductive logic. It is a sad commentary on our times that we need to make explicit this proviso.

$$
\mathbf{R}\begin{pmatrix} A_{\kappa=0}(0) \\ B_{\kappa=0}(0) \\ C_{\kappa=0}(0) \\ D_{\kappa=0}(0) \end{pmatrix} = \begin{pmatrix} 1 & 1 & 1 & -1 & -1 & -1 & -1 & 1 \\ 1 & 1 & -1 & 1 & -1 & -1 & 1 & -1 \\ 1 & -1 & 1 & 1 & -1 & 1 & -1 & -1 \\ -1 & 1 & 1 & 1 & 1 & -1 & -1 & -1 \end{pmatrix} \equiv \mathbf{R}_{-1}. \quad (i)
$$

Here I am denoting the specific experimental quantity observation vector in its shorthand form using the subscript κ just mentioned: $[A_{\kappa=0}(0), B_{\kappa=0}(0), C_{\kappa=0}(0), D_{\kappa=0}(0)]^T$. The subscript $\kappa = 0$ on each of the spin observation values denotes the contextual value of the angle combination pertaining to the experiment, and the arguments of the vector components designate that each of the four angles ϕ_i is equal to 0 radians. However, this same realm matrix of possibilities would pertain to a spin observation vector resulting from *any* design that meets this four-angle condition, $\kappa = 0$.

Since the restriction $\phi_1 + \phi_2 - \phi_3 - \phi_4 = 0$ implies via equation (9) that the *expected* spin-product $E^{\psi}[ABCD(\phi_1, \phi_2, \phi_3, \phi_4)]$ equals -1, it would be impossible to achieve any four experimental spin results that allow the vector of these multiplicands to imply a positive product. The quantum probability weight on such observation vectors must equal 0, and such observation vectors would be impossible. (If such an observation *were* seemingly made, the experimental quantum theorist would typically reject its validity, and check the settings of the direction angles of the four Stern-Gerlach magnets in the experimental run that generated it.) In the entangled state of the four-particle system specified by $|\Psi\rangle$ in equation (7), the columns of this matrix exhaust the vector values of measurements that can arise from such an experiment. The product of any of these eight ensemble column components equals -1.

Notice the concluding *definition symbol*, ($\equiv$), introducing the denotation $\mathbf{R}_{-1}$ at the right end of equation (*i*). This is to distinguish it from a companion matrix to be denoted by $\mathbf{R}_{+1}$, which specifies the realm matrix corresponding to the possible outcomes of a *different experiment* in which the magnet angles satisfy instead the alternative condition providing for (11*b*), that $\phi_1+\phi_2-\phi_3-\phi_4 = \pi$. Again, without loss of generality, an exemplar experiment would generate an observable result $(A_{\kappa=\pi}(\pi), B_{\kappa=\pi}(0), C_{\kappa=\pi}(0), D_{\kappa=\pi}(0))^T$ with a realm matrix of columns whose products all equal $+1$:

$$\mathbf{R}\begin{pmatrix} A_{\kappa=\pi}(\pi) \\ B_{\kappa=\pi}(0) \\ C_{\kappa=\pi}(0) \\ D_{\kappa=\pi}(0) \end{pmatrix} = \begin{pmatrix} 1 & -1 & -1 & -1 & 1 & 1 & 1 & -1 \\ 1 & -1 & 1 & 1 & -1 & -1 & 1 & -1 \\ 1 & 1 & -1 & 1 & -1 & 1 & -1 & -1 \\ 1 & 1 & 1 & -1 & 1 & -1 & -1 & -1 \end{pmatrix} \equiv \mathbf{R}_{+1} . \quad (ii)$$

The restrictions on the measurement vector possibilities embedded in the realm matrices $\mathbf{R}_{-1}$ and $\mathbf{R}_{+1}$ derive from equation (9) of GHSZ, which specifies that

$$E^{\psi}(\hat{\mathbf{n}}_1, \hat{\mathbf{n}}_2, \hat{\mathbf{n}}_3, \hat{\mathbf{n}}_4) = -cos(\phi_1 + \phi_2 - \phi_3 - \phi_4).$$

At the two extreme angle restrictions we have entertained, that $\phi_1 + \phi_2 - \phi_3 - \phi_4$ equal 0 or π, this negative cosine value equals -1 and $+1$ respectively. This is what restricts the measurement realms to be $\mathbf{R}_{-1}$ and $\mathbf{R}_{+1}$ in these extreme cases. If the combination of experiment angles in this equation were to equal some other value of $\kappa \in (0, \pi)$, then the realm matrix of possibilities for the measurements of the four electron spins would be the concatenation of these two realms, $[\mathbf{R}_{-1} \ \mathbf{R}_{+1}]$. With the recognition of this situation clearly in mind, we are ready to face the wall.

Hitting a wall at equation (16) ? Do we get there ?

GHSZ regard their conclusion (16), that $A_\lambda(2\phi) = A_\lambda(0) = constant$ for any angle ϕ as surprising, for reasons they have well explained. Thinking, nonetheless, that this equation is not mathematically contradictory in itself, they were sidetracked into a mistaken analysis that presumes jointly the contradictory suppositions of their full two-lined expressions of $(11a)$ and $(11b)$. "Finding" a contradiction in their results, they quit.

Having recognised their error in presuming contradictory suppositions, we shall now continue to investigate the apparent conundrum which they ignored. We have already recognised that their equation (16) derives from analysis that uses incomplete notation for describing the situation. We shall now attempt to derive their "troublesome" result, using the expanded notation I have suggested. I have embellished their statement of equations $(11a, b)$ and $(12a, b, c, d)$ in the developments below to the formulations in $(11^*a, b)$ and $(12^*a, b, c, d)$. Let's now try to follow their line of argument using our complete notation, and see whether we will still hit the wall at their (16).

Heading toward the wall

As a consequence of equalities $(12a)$ and $(12b)$, GHSZ obtained

$$A(0)B(0)C(0)D(0) \;=\; A(\phi)B(0)C(\phi)D(0)\,,$$

which allowed them to conclude

$$A(\phi)C(\phi) \;=\; A(0)C(0)$$

through cancellation of $B(0)D(0)$ which appears identically on both sides of this equation. However, we find that writing the equations $(12a, b)$ in full notation which includes the contextual subscripts, the equalities (12^*a) and (12^*b) obtain for us an equation that looks quite different:

$$A_{(0,0,0,0)}(0)B_{(0,0,0,0)}(0)C_{(0,0,0,0)}(0)D_{(0,0,0,0)}(0)$$

$$= \; A_{(\phi,0,\phi,0)}(\phi)B_{(\phi,0,\phi,0)}(0)C_{(\phi,0,\phi,0)}(\phi)D_{(\phi,0,\phi,0)}(0)\,. \quad (13^*a)$$

In full notation, the multiplicands that appeared identically as $B(0)D(0)$ on the left-hand sides of $(12a)$ and $(12b)$, now appear as two quite different things in (12^*a) and (12^*b), ... as they should!

For they represent the products of B and D spins observed in two different contexts: $B_{(0,0,0,0)}(0)D_{(0,0,0,0)}(0)$ and $B_{(\phi,0,\phi,0)}(0)D_{(\phi,0,\phi,0)}(0)$. In fact, *every* multiplicand observation on either side of equation (13^*a) is distinct from its counterpart on the other side. The consequence of this recognition will require that we examine the implications of presuming the principle of local realism for the situation. This was the supposition insisted by Einstein in the EPR paper, which stimulated the considerations of Bell's inequality from its inception. Yes, we will require some more thinking.

How might local realism be relevant?

It is widely understood and generally accepted among quantum theorists that the experimental outcomes of paired particle emissions in a quantum experiment depend stochastically on the structural design of the setup at both observation stations. This phenomenon is commonly referred to as "the entanglement of quantum particles" (a nomenclature we shall find reason to suspect in a later chapter). We learned in Chapter 1 how this applies to a pair of photons in Aspect's experiment, and it applies as well to the *two* pairs of electrons emitted in the GHSZ experiment. This is motivated in this context by the quantum theoretic result that $E^{\psi}[ABCD(\phi_1, \phi_2, \phi_3, \phi_4)] = -\cos(\phi_1 + \phi_2 - \phi_3 - \phi_4)$. The associated implication is that the observed spin-product value at any station depends stochastically on the magnet angles at all four stations. This awareness underlies our current insistence on the expressed designation of the spin observation at station B, say, as $B_{\phi_1,\phi_2,\phi_3,\phi_4}(\phi_2)$.

Einstein concurred with the probabilistic assessments of quantum observations. However, he claimed that the stochastic character of the propositions derives from our uncertain knowledge of the situation rather than from a stochastic feature of the particles themselves. The theory must be incomplete in its ordinations, because if it were not then the results would purport to support "spooky action at a distance". This would amount to the transmission of information about the magnet directions from station to station at a speed exceeding the speed of light, something the theory of relativity does not countenance. (The magnet directions at the observation stations are not determined until after the electrons have left their source.)

Although admittedly the quantum probabilistic assessment of an experimental outcome at one station depends on the setting of the experiment at another station, the "principle of local realism" would qualify this understanding. Local realism would provide that the result of a gedankenexperiment on a particle observed in a specific instance of the setup would necessarily remain the same if it were also observed in this instance in the same way locally while the setup of the experiment at another station differed. For locally, the experimental setup would be the same. This proclamation, outside the realm of quantum theory itself (which proclaims nothing about the results of two alternative experimental designs that cannot be jointly instantiated) is relevant to an analysis of equation (13^*a) to which we now return for consideration.

In a gedankenexperiment on a single quartet of electrons, there are some specific equalities among designated components of equations (12^*) that would derive from a presumption of local realism. This says that the instantiated numerical value of spin activity in specific location conditions, such as might be identified by $B_{0,0,0,0}(0)D_{0,0,0,0}(0)$, should be identical to its value in any other instance of the same *local* conditions. It should not matter what might be occurring simultaneously at other stations *outside of this locality*, as long as the value of $\kappa = \phi_1 + \phi_2 - \phi_3 - \phi_4$ entailed remains 0. An alternative corresponding exemplar would be designated by $B_{\phi,0,\phi,0}(0)D_{\phi,0,\phi,0}(0)$ for some value of ϕ. This alternative setup supports the same value of κ, and thus the same value of expected spin-product. Simultaneous assessment of both of these product values requires the consideration of a gedankenexperiment. We must be careful.

What does local realism imply, algebraically?

In the first place, the principle of local realism explicitly requires the equality of two pairs of terms that appear in equations (12^*a) and (12^*b). These are

$$B_{0,0,0,0}(0) = B_{\phi,0,\phi,0}(0) \text{ and } D_{0,0,0,0}(0) = D_{\phi,0,\phi,0}(0) . \quad (14)$$

In all four subscripted magnet configurations, the contextual combination of magnet directions sets $\kappa = 0$. This distinguishes the quantum condition specified by GHSZ in equation $(11a)$ from that of equation $(11b)$, repeated in our completed notation as (11^*a)

and (11*b). Both equations (12*a) and (12*b) pertain, supporting a 4-spin-product value of -1 in each case. Local realism insists on equality of the two particular pairs of B and D components in a gedankenexperiment on the same pair of electrons in the two settings, identical no matter what would have been the settings of magnet angles for the spins A and C, nor the spin values there. Of course, the alternative magnet angles at A and C would need be equal as well, as seen in the first and third subscripted designations of $B_{\phi,0,\phi,0}(0)$ and $D_{\phi,0,\phi,0}(0)$, so as to instantiate $\kappa = 0$.

Do notice, however, that the two equalities of (14) could be satisfied by any possible vector of spin possibilities that meets the condition $\kappa = 0$ on the magnet settings. It is evident from matrix $\mathbf{R}_{-1}$ of equation (i) that either, both, or neither of the values of the components $B_{0,0,0,0}(0)$ and $D_{0,0,0,0}(0)$ could be positive (or negative) in the subscription to local realism. So might be *their* product, which is only a sub-product of the product of all four spin values. It is only this *product of all four* that must equal -1.

With this awareness in mind, take a look again at the equations designated (12*a) and (12*b) using complete explicit notation:

$$A_{0,0,0,0}(0)B_{0,0,0,0}(0)C_{0,0,0,0}(0)D_{0,0,0,0}(0) \;=\; -1$$

$$A_{\phi,0,\phi,0}(\phi)B_{\phi,0,\phi,0}(0)C_{\phi,0,\phi,0}(\phi)D_{\phi,0,\phi,0}(0) \;=\; -1$$

While presuming local realism, we can surely divide both of the equi-valued left-hand sides of these equations by the corresponding sides of the product equation shown in (14). These insist that $B_{0,0,0,0}(0)D_{0,0,0,0}(0) = B_{\phi,0,\phi,0}(0)D_{\phi,0,\phi,0}(0)$. Of course, this yields a result that

$$A_{0,0,0,0}(0)C_{0,0,0,0}(0) \;=\; A_{\phi,0,\phi,0}(\phi)C_{\phi,0,\phi,0}(\phi) \,. \qquad (15)$$

However, we need recognise that the numerical value of these resulting equal products might well be either -1 or $+1$. No column of the realm matrix of possible vector outcomes of the spin measurements displayed in equation (i) is proscribed by these considerations. Thus, the equal value of the products $B_{0,0,0,0}(0)D_{0,0,0,0}(0)$ and $B_{\phi,0,\phi,0}(0)D_{\phi,0,\phi,0}(0)$ by which we had divided the left-hand sides of equations (12*a) and (12*b) might be either -1 or $+1$.

Now condition (15) is admittedly a novel recognition, because it does not arise directly in the stated principle of local realism pertinent to the spin values of B and D. This new implied condition (15) specifies the equality of two different spin-products under conditions that are *not identical at their own stations*. It arises in contexts in which locality does pertain directly when the angles at stations B and D are unchanged. We are now assessing the equality of the product of spin values AC in conditions of two situations when the directions of their magnetic sensors are different. Yes, this *is* consequential. Nonetheless, it is evident that either, neither, or both component values of the A and C observations might equal either $+1$ or -1, even while subscribing to the local realism equation (14). For not a single one of the observation possibilities for the gedanken spin vector displayed in the realm matrix $\mathbf{R}_{-1}$ has been eliminated from contention as an experimental result by the considerations we have raised. Examine it once again, above.

To understand the implications of this awareness for the claims of GHSZ, we need now to consider their continuing derivation which involves the simultaneous joint implications of equations (12^*a) and (12^*c). A similar analysis of these two equations would yield the parallel result that

$$A_{0,0,0,0}(0)D_{0,0,0,0}(0) = A_{\phi,0,0,\phi}(\phi)D_{\phi,0,0,\phi}(\phi), \qquad (16)$$

again allowing that the two sides of this equality may equal either -1 or $+1$. For again, this prescription too could be satisfied by any column of the realm matrix $\mathbf{R}_{-1}$.

Pursuing the error to absurdity

In designating the component spin values of observations without the subscripted specification of context, the casual notation of GHSZ characterises our conclusions (15) and (16) as merely $A(0)C(0) = A(\phi)C(\phi)$ and $A(0)D(0) = A(\phi)D(\phi)$. Dividing the two sides of the first equation by the ostensibly identical sides of the second, they arrive at their substantive conclusion that $C(0)/D(0) = C(\phi)/D(\phi)$, ... and thus $C(0)D(0) = C(\phi)D(\phi)$, since both $D(0)$ and $D(\phi)$ can equal only -1 or $+1$. Continuing then, *still oblivious to contextual subscripts,* they would be replacing the value of $C_{(2\phi,0,\phi,\phi)}(\phi)D_{(2\phi,0,\phi,\phi)}(\phi)$ in $(12d^*)$ with $C_{(0,0,0,0)}(0)D_{(0,0,0,0)}(0)$, a completely different animal! Then they cancel the apparent triple product in the resulting left-hand side form of their $(12d)$, which appears to them merely as $B(0)C(0)D(0)$, with the ostensibly identical terms appearing in $(12a)$. This yields them their really strange result that $A(2\phi) = A(0)$ *for any magnet angle* ϕ. They are astonished particularly by their "troublesome" result that $A(\pi) = A(0)$, which would pertain when $\phi = \pi/2$. It is their replacement of the values of $C(\phi)D(\phi)$ in $(12d)$ with $C(0)D(0)$ that is not warranted.

While decrying the content of their result that $A(2\phi) = A(0)$ as unusual on account of its defying our experience with directions of magnetic force, they ignore a more stunning implication of their "derivation": that apparently the value of a spin observation A must be identical no matter at what angle is directed the spin-detecting Stern-Gerlach magnet at the station, and no matter what the contextual directions of the other four magnets. *This result is not puzzling at all. It is simply gibberish:* JUST PLAIN WRONG. We never reach the wall at all.

In contrast, an appropriate assessment of equations (15) and (16) is that these could be substantiated in accordance with local realism by any column of the possibility matrix (i) recognised by quantum theory. Using contextual notation for individual spin values, it is clear that the numerical values of the left-hand sides of equations (12^*a) and (12^*b) can be identical, even while the sub-products displayed by $B_{0,0,0,0}(0)D_{0,0,0,0}(0) = B_{\phi,0,\phi,0}(0)D_{\phi,0,\phi,0}(0)$ may equal either -1 or $+1$ in the columns allowed by $\mathbf{R}_{-1}$. Corresponding to whichever result occurs in an observed column would be a value of $A_{0,0,0,0}(0)C_{0,0,0,0}(0) = A_{\phi,0,\phi,0}(\phi)C_{\phi,0,\phi,0}(\phi)$ equal to the negative of the BD product in that column. These realisations allow us to conclude that the restriction of local realism in a gedankenexperiment would allow either, neither, or both of the terms $C_{\phi,0,\phi,0}(\phi)$ and $D_{\phi,0,0,\phi}(\phi)$ to equal -1 or $+1$. The wanton cancellations in the machinations of GHSZ to yield their troublesome result are not afforded by the results of quantum theory.

The puzzle of the puzzling result has dissolved, along with the purported contradiction involved in the formulation of the principle of local realism in dimensions greater than two.

Where have we arrived?   Recognition of symmetry

Well, what is to be made of the considerations of GHSZ? We had followed them without concern through their equations (9) and (10), notwithstanding our allusion to their use of cryptic notation. We had proposed that the conclusion of their analysis would be expressed more clearly as
$$E^{\psi}[ABCD(\phi_1, \phi_2, \phi_3, \phi_4)] = -cos(\phi_1 + \phi_2 - \phi_3 - \phi_4).$$
Using this appropriately expressive notation, the entanglement among the four multiplicand spins is evident. Moreover, it is readily apparent that this conclusion displays a symmetry of the quantum spin-product expectation with respect to rotations of the full 4-ply

67

Stern-Gerlach mechanism in the (x, y) dimensions at the four observation stations, and more! For as long as $t_1 + t_2 = t_3 + t_4$, it is evident that

$$E^\psi[ABCD(\phi_1 + t_1, \phi_2 + t_2, \phi_3 + t_3, \phi_4 + t_4)]$$
$$= E^\psi[ABCD(\phi_1, \phi_2, \phi_3, \phi_4)], \qquad (iii)$$

because the value of κ associated with both of these 4-angle designs is identical. Let's continue to think!

Both rotational and permutation symmetry

Algebraically, two types of symmetry are evident in the stipulations of equations (8) and (9) motivated by quantum theory. These show that the expected 4-spin-product is a function only of the angle combination $\phi_1 + \phi_2 - \phi_3 - \phi_4$. Rotational symmetry of the experimental conditions would be exhibited in the transformation of the vector of angles $(\phi_1, \phi_2, \phi_3, \phi_4)$ by the addition of any vector of constants (t, t, t, t), which would surely preserve the angle combination κ. In fact, preservation would continue under the addition of any angle vector $\mathbf{t}_4$ for which $t_1 + t_2 = t_3 + t_4$. This general condition allows permutations of either or both of ϕ_1 with ϕ_2 and/or of ϕ_3 with ϕ_4 in the specification of κ. It would also allow the permutation of any pair (ϕ_1, ϕ_2) with (ϕ_3, ϕ_4), because the cosine of any angle is identical to the cosine of the negative of that angle.

Geometrically, simple rotational symmetry would allow rotation of the (x, y) planes containing the directional vectors $(\hat{\mathbf{n}}_1, \hat{\mathbf{n}}_2, \hat{\mathbf{n}}_3, \hat{\mathbf{n}}_4)$ in Figure 1 around the (x, y) axes orthogonal to the $^-\mathbf{z} \leftrightarrow {}^+\mathbf{z}$ axes, all to the same degree. Permutation symmetry would allow the exchange of directional designations of (x, y) axes between stations 1 and 2 or between 3 and 4, ... or an exchange of these axis systems of stations $(1, 2)$ with those of $(3, 4)$. Moreover, the invariance with respect to $t_1 + t_2 = t_3 + t_4$ is even richer than all these symmetries. It allows the twisting of the (x, y) axes at stations (1) and (2) in any ways one wishes, just so long as one would twist the axes at stations (3) and (4) correspondingly, so to ensure that the *sum* of the two latter twisting degrees equals the *sum* of the twisting degrees of the the former two.

A final comment on this situation is in order before we consider a Bell inequality formulation in the context of the 4-dimensional GHSZ experiment proposal. Notice that the QM-motivated invariance equation (iii) atop this page specifies the invariance of the

68

expectation of *the product* of the four spin observations with respect to all these forms of twisting any initial reference 4-angle configuration. However, equation *(iii)* does *not* specify invariance of the component multiplicands themselves that generate the product. Each time we suggest a twisting of the four Stern-Gerlach magnet orientations in the (x, y) planes at the four stations, however we propose to do it, we are considering a new distinct run of the 4-ply spin experiment of GHSZ. Even in the special cases that the initial reference configuration specifies κ equal to 0 or to π, the invariance pertains to the product of the multiplicand spin observations, not to the multiplicand spin values themselves.

Now why did GHSZ propose their analysis in the first place? They were attempting to study a higher dimensional example of the Bell gedankenexperiment, to see how the inequality might fare. Before they were deterred and sidetracked from this concern by their errors of logic and of casual notation, they were trying to formalise the implications of the principle of local realism for a four-dimensional experiment. As it turns out, the formalisation *can* be constructed without confronting any contradiction at all, but we must be a little bit careful. A preliminary airing of a few issues will conclude my exposition in this chapter.

Extending CHSH/Bell to four dimensions

Remember that Einstein's principle of local realism pertains to situations that lie outside the scope of quantum theory. In the context of a GHSZ 4-ply experiment on two pairs of electrons, his assertion concerns the outcome of other experiments that one might imagine conducting in tandem with the 4-ply experiment that one does conduct. However, these others are precluded from concomitant execution by the one that is conducted. According to the general form of the uncertainty principle, quantum theory expressly renounces any scope for making proclamations about the joint result of such incompatible ventures. Of course it can and does make probabilistic statements separately about the results of any one of the 4-ply experiments that might be considered. No worries. We can at least *think* about the prospect of such an impossible joint venture, and assess the relevance to its conduct of what QM does say. Many, many people have.

Well, what would we need to think about if we were to think about such vagaries in the context of the GHSZ experiment? At least we now have a vocabulary and a syntax of language to talk about what we are thinking.

The principle of local realism specifies that while the results of a quantum experiment may well be stochastic, assigned various probabilities according to the tenets of quantum theory, if the results at the observation stations 1 and 3 in any actual experimental instance equal $+1$ and -1, say, then these results would have had to be the same in this instance at other specific circumstances of their companion angles as well. Let's see what this means.

Suppose, for example, that the outcome of a specific run of the experiment characterised by a 4-angles setting of $(\phi_1, \phi_2, \phi_3, \phi_4)$ for which $\kappa = \phi_1 + \phi_2 - \phi_3 - \phi_4 = 0$, yields the observed spin results

$$[A_{(\phi_1,\phi_2,\phi_3,\phi_4)}(\phi_1), B_{(\phi_1,\phi_2,\phi_3,\phi_4)}(\phi_2), C_{(\phi_1,\phi_2,\phi_3,\phi_4)}(\phi_3), D_{(\phi_1,\phi_2,\phi_3,\phi_4)}(\phi_4)]$$
$$= (+1, -1, -1, -1) . \qquad (17)$$

Notice that the spin-product of the components of this supposed observation vector is -1, as is required for a setup specifying this value of $\kappa = 0$. About such an observation, the principle of local realism would say, among other things, that if we were to conduct concomitantly a companion experiment *on this same quartet of electrons* at adjusted angle settings of the form $(\phi_1, \phi_2 + t, \phi_3, \phi_4 + t)$, preserving the value of $\kappa = 0$, then the result vector for this alternative experiment would only have to satisfy the form

$$[A_{(\phi_1,\phi_2+t,\phi_3,\phi_4+t)}(\phi_1), B_{(\phi_1,\phi_2+t,\phi_3,\phi_4+t)}(\phi_2 + t),$$
$$C_{(\phi_1,\phi_2+t,\phi_3,\phi_4+t)}(\phi_3), D_{(\phi_1,\phi_2+t,\phi_3,\phi_4+t)}(\phi_4 + t)]$$
$$= (+1, x, -1, x). \qquad (18)$$

While A and C must remain $(+1, -1)$, the value of x common to B and D might equal either -1 or $+1$. Either value would ensure that the product of the component observations remains equal to -1 as required. The claim of local realism is that the behaviour of these electrons at stations 1 and 3 would remain the same in any other run in any other setting of the angles at stations 2 and 4, and moreover that the prescriptions of quantum probabilities must also be preserved.

Even if the twists of the magnets at 2 and 4 would destroy the value of κ in the concomitant run (via different specifications of the t's at these stations) then the spin values at 1 and 3 *for this same pair of electrons* should still remain the same according to local realism, but the concomitant spin values at 2 and 4 might enjoy more liberty. This would depend on the value of κ associated with the adjusted magnet angles $\phi_2' = \phi_2 + t_2$ and $\phi_4' = \phi_4 + t_4$. If the value of κ for the adjusted magnet angles were not equal to either 0 or π, then the spin values at stations 2 and 4 would be permitted to alternate in sign as well, in any way they might. This would be allowed because the joint distribution of the component observations (A, B, C, D) would be different in the adjusted angle settings according to quantum theory. Even the QM expectation of the product $ABCD$ would then be different as well. It would be only the spin values of the two electrons at the unchanged magnet angles that local realism would require to remain the same in two concomitant runs on the same pairs of electrons.

It is because the electrons are entangled that specification of the consequences of local realism must take all four angle settings into account in the higher dimensional context. Both of these distinct spin vector observations $(1, 1, -1, 1)$ and $(1, -1, -1, -1)$ would ensure that the product of the spin values would equal -1 when the angles ϕ_2 and ϕ_4 are both twisted by the same incremental value t in the scenario of (18). So there are two possible observation vectors for this concomitant experiment that would satisfy this proclamation of local realism, though there are many more impossibilities that would not. For examples, an observation vector $(1, 1, -1, -1)$ would be impossible in the twisted case, since it does not even respect the tenets of quantum theory. (Remember we are presuming that $\kappa = \phi_1 + \phi_2 + \phi_3 - \phi_4 = 0$, so the product of the four observations must equal -1.) Further, an observation of $(1, -1, 1, 1)$ would be impossible as well, not being in accord with local realism since the value of $C_{(\phi_1, \phi_2, \phi_3, \phi_4)}(\phi_3)$ would have changed.

I shall now outline the construction of a Bell-type inequality in CHSH form that would be appropriate to the four-dimensional problem of GHSZ, delineating relevant details showing how this might be done. We shall find that neither is there anything contradictory about it, nor would quantum theoretic probabilities defy it. As befits the pleasures of quantum physical theorising, there will be intrigue involved. What I propose is to design the performance of a GHSZ gedankenexperiment on a single quartet of doubly paired electrons under 16 distinct conditions. These amount to all combinations of two magnet angle settings at each station, ϕ_i and ϕ_i'.

We shall consider specific paired magnet orientations at the opposing stations 1 and 3 and at the opposing stations 2 and 4 that mimic the paired polarisation angles used by Aspect in his experiments with single pairs of photons. The experiments at stations 1 and 3 on their own (ignoring the parallel experiments at stations 2 and 4) would replicate the simpler experiments of Aspect/Bell at stations A and B. The structure is similar for the pairings at stations 2 and 4, ignoring those at 1 and 3. Of course, in the GHSZ context which we continue, the algebraic details apply to electron spins rather than to photon polarisations. However, as is well known, the pertinence of issues relevant to Bell's inequality are equivalent.

Implementing the extended gedankensetup

We can now proceed with a description of a gedankenexperiment in which the two pairs of electrons are sent to the four stations, concomitantly toward all possible choices of two different magnet angles at each station. This procedure would involve sixteen such angle configurations. I shall be brief, merely reporting the results of the analysis at each stage of its development rather than providing every interim detail.

The two chosen angles at each station i will be designated as ϕ_i and ϕ_i', for $i = 1, 2, 3$, and 4. According to the principle of local realism, we presume that any specific instantiation of a spin observation at a station at a specific angle would be identical, no matter which other three angles were engaged by the electrons at the other stations. This is despite the fact that the joint quantum probability distribution for the outcome of the vector of all four observations (and for their product) at a specific 4-angle arrangement clearly depends on the settings of all four magnet angles. This is exactly the principle we followed in generating results of the simpler gedankenexperiment on single photon pair in the framework of Aspect/Bell. In that experiment on photon pairs, there were four distinct angles between the two polariser directions at the two observation stations, involving $2^4 = 16$ possible vectors of four polarisation results. The corresponding design of a GHSZ gedankenexperiment would involve eight distinct magnet directions, two for each of the four observations stations, specifying $2^8 = 256$ possible vectors of spin results that might arise. Each vector would identify a spin result for two magnet directions at each of the four observation stations.

Designing the GHSZ magnet directions ϕ_i and ϕ'_i at the opposing paired stations $(1, 3)$ and $(2, 4)$ to correspond to the polarisation directions employed in Aspect's experiments would involve specifications of $(\phi_1, \phi'_1) = (\phi_2, \phi'_2) = (7\pi/16, 3\pi/16)$, and $(\phi_3, \phi'_3) = (\phi_4, \phi'_4) = (5\pi/16, 1\pi/16)$. These settings would make the four *relative angles* between the magnet directions from station 1 to station 3 equal to $-\pi/8, -3\pi/8, \pi/8$ and $-\pi/8$. (These angles arrise from the differences $\phi_3 - \phi_1, \phi'_3 - \phi_1, \phi_3 - \phi'_1$, and $\phi'_3 - \phi'_1$.) These differences are identical to the relative angles between polarisers employed by Aspect (as in our Chapter 1) as $(\mathbf{a}, \mathbf{b}), (\mathbf{a}, \mathbf{b}'), (\mathbf{a}', \mathbf{b})$, and $(\mathbf{a}', \mathbf{b}')$ in his experiment with a single pair of photons. The same would characterise the relative angles between the magnet directions at stations 2 to 4. Aspect's polariser directions themselves, $\mathbf{a}, \mathbf{b}, \mathbf{a}'$ and $\mathbf{b}'$, would correspond to the magnet directions ϕ_1, ϕ_3, ϕ'_1 and ϕ'_3, for example.

The scale of the gedankenexperiment has grown exponentially. Aspect/Bell dealt with $2^2 = 4$ simultaneous runs of polarisation experiments involving none, one, or two primed direction settings of the two station polarisers. There are $^2C_0 = 1, ^2C_1 = 2$, and $^2C_2 = 1$ of these. The GHSZ design involves $2^4 = 16$ simultaneous runs of spin experiments involving none, one, two, three, or four primed values of magnet directions, making $^4C_0 = 1, ^4C_1 = 4, ^4C_2 = 6, ^4C_3 = 4$, and $^4C_4 = 1$ of these: that is, 16 in all.

Identifying now the realm matrix of possible observations for the eight electron spin results involved in such a sixteen-fold gedankenexperiment, we would describe it in transposed vector form as

$$\mathbf{R}[A(\phi_1), B(\phi_2), C(\phi_3), D(\phi_4), A(\phi'_1), B(\phi'_2), C(\phi'_3), D(\phi'_4)]^T,$$

and would recognise the matrix columns as composing the cartesian product of eight component vectors, $\{-1, +1\}^8$. Comparably in the Aspect/Bell gedankenexperiment on single pair of photons, we constructed the realm matrix $\mathbf{R}[A(\mathbf{a}), B(\mathbf{b}), A(\mathbf{a}'), B(\mathbf{b}')]^T$ and recognised the matrix columns as composing the cartesian product of four component vectors, $\{-1, +1\}^4$.

Beneath each column of this 8×256 matrix, we would compute the vector of 4-spin-products corresponding to this possibility for the *sixteen* scenarios of concomitant gedankenexperiments. This vector would designate the ordered 4-spin-products using none, one, two, three, and then four of the "alternate" primed magnet angles ϕ'_i rather than the base orientation at angle ϕ_i. The base angles are retained at the others.

Without printing the full 16×256 elements of the top section of the realm matrix, the schema below lists in its first column the names of the sixteen 4-spin-product quantities each column of the realm would generate. In the second and third columns appear the value of κ that each would engender, and the quantum Expectation of the spin product when the magnet angles are set as designated: $(\phi_1, \phi_1') = (\phi_2, \phi_2') = (7\pi/16, 3\pi/16)$, and $(\phi_3, \phi_3') = (\phi_4, \phi_4') = (5\pi/16, 1\pi/16)$. The value of κ for each spin-product configuration equals $\phi_1^* + \phi_2^* - \phi_3^* - \phi_4^*$, and the QM-motivated expectations for these 4-spin-products are

$$E[A(\phi_1^*)B(\phi_2^*)C(\phi_3^*)D(\phi_4^*)] = -\cos(\phi_1^*+\phi_2^*-\phi_3^*-\phi_4^*) = -\cos(\kappa^*),$$

where each ϕ_i^* corresponds to either ϕ_i or ϕ_i' as is appropriate.

Spin-products involved in a GHSZ Gedankenexperiment

Four-product-designation	κ Value	Expectation
$\mathbf{R}[A(\phi_1)B(\phi_2)C(\phi_3)D(\phi_4)$	$\pi/4$	$-1/\sqrt{2}$
$A(\phi_1')B(\phi_2)C(\phi_3)D(\phi_4)$	0	-1
$A(\phi_1)B(\phi_2')C(\phi_3)D(\phi_4)$	0	-1
$A(\phi_1)B(\phi_2)C(\phi_3')D(\phi_4)$	$\pi/2$	0
$A(\phi_1)B(\phi_2)C(\phi_3)D(\phi_4')$	$\pi/2$	0
$A(\phi_1')B(\phi_2')C(\phi_3)D(\phi_4)$	$-\pi/4$	$-1/\sqrt{2}$
$A(\phi_1')B(\phi_2)C(\phi_3')D(\phi_4)$	$\pi/4$	$-1/\sqrt{2}$
$A(\phi_1')B(\phi_2)C(\phi_3)D(\phi_4')$	$\pi/4$	$-1/\sqrt{2}$
$A(\phi_1)B(\phi_2')C(\phi_3')D(\phi_4)$, for which $\kappa = \pi/4$ and $E = -1/\sqrt{2}$		
$A(\phi_1)B(\phi_2')C(\phi_3)D(\phi_4')$	$\pi/4$	$-1/\sqrt{2}$
$A(\phi_1)B(\phi_2)C(\phi_3')D(\phi_4')$	$3\pi/4$	$+1/\sqrt{2}$
$A(\phi_1')B(\phi_2')C(\phi_3')D(\phi_4)$	0	-1
$A(\phi_1')B(\phi_2')C(\phi_3)D(\phi_4')$	0	-1
$A(\phi_1')B(\phi_2)C(\phi_3')D(\phi_4')$	$\pi/2$	0
$A(\phi_1)B(\phi_2')C(\phi_3')D(\phi_4')$	$\pi/2$	0
$A(\phi_1')B(\phi_2')C(\phi_3')D(\phi_4')]$	$\pi/4$	$-1/\sqrt{2}$.

Having studied the simpler problem of Aspect/Bell, it is now no surprise to find that the realm matrix so generated for the sixteen 4-spin-products does not display 256 distinct columns, but rather merely 32. Many columns are duplicates. Moreover, the column rank of the reduced matrix of distinct columns is only 5. What this tells us, is that if we can identify any five rows of this 16×32 matrix whose columns exhaust the cartesian product $\{-1, +1\}^5$ then the remaining rows will have been identified to be functionally related to them. It turns out that there are many such combinations of five rows that will do this. Of the $^{16}C_5 = 4368$ available for such choices of five rows, 2688 of them are found to provide such a functional relation. In fact, all of these functional relations hold among the sixteen components of any column vector in the realm we have just designated. Although this is quite a step up in the complexity of the functional restrictions among the 4-spin-products of a gedanken-experiment, it is structurally no different than the four simultaneous functional relations that bound up the polarisation products in the simpler context of the eight dimensional Aspect/Bell problem.

Let's stop for a moment to think about what this means: the conditional distribution for the other eleven 4-spin-products given any of these five is degenerate on the function values they stipulate. Ostensibly, we would be searching for a QM-motivated joint probability distribution over the possible results of all sixteen 4-spin-product quantities in the GHSZ gedankenexperimental design. These constitute the thirty-two columns of our realm matrix. Of course, the general uncertainty principle tells us that there is no such joint distribution, because quantum theory says nothing about the joint result of any incommensurable 4-products. However, if there were such a distribution, then it could be factored as the product of the marginal joint distribution for any five of them times the conditional distribution for the other eleven given these five. Since this conditional distribution relative to any of the 2688 function-producing choice of five from sixteen is degenerate, the joint distribution for all sixteen would resolve to a specification of the joint distribution for these five. Again however, quantum theory does not uniquely identify such a distribution.

What quantum theory does do is to specify the expectation of the product of spin values that might be observed in a GHSZ experiment in any single 4-angle setting. These could be assessed for any five function-generating 4-spin products to determine bounds on the joint distribution for all sixteen of them. Along with the the summation constraint, the five such product expectation values would place six linear constraints on the probabilities for the thirty-two constituents of the partition underlying the joint distribution. It is precisely this situation to which the computational structure of linear programming problems applies. Structurally, it is identical to the problem we have formulated to determine bounds on the expected combination of polarisation products for the conundrum of Aspect/Bell.

Well, what should be the objective function of such a linear programming problem? We have seen that the Bell quantity "s" in the Aspect/Bell analysis of the CHSH formulation is a linear combination of the Expected products of all four components of the gedankenexperiment involved. The coefficients identifying the linear combination are all either $+1$ or -1. We might consider similar linear combinations of the sixteen 4-spin-products as quantities of interest in the experiment on the double electron pairs. GHSZ never got around to assessing this situation. There are 2^{16} such combinations they might wish to consider as their objective function, with coefficients exclusively of -1 and $+1$. Whatever combination might be proposed, five linear constraints on the spin-products for the function domain settings, plus the summation constraint, would leave 26 dimensions of freedom in the argument vector of the objective function. The linear programming computation would find the vectors $\mathbf{q}_{32}^{min}$ and $\mathbf{q}_{32}^{max}$ that yield the extreme values of this linear combination.

To compute a complete solution to such a gedanken problem, we would need to compute 2688 pairs of LP problems (*min* and *max*) to find all the vertices of the polytope of probability distributions that quantum theory would allow as its solution. Of course, as in the corresponding Aspect/Bell solution, many of these would be duplicates. The Bell problem was much simpler, involving only four choices of functional relations that determine the fourth polarisation product from three of them. The GHSZ problem involves 2688 functional relations that generate eleven 4-spin-products from five. At any rate, we would arrive at a larger sized solution of a quantum-theory-specified polytope of probability distributions. Structurally, it would be similar to the solution we found in the case of Aspect/Bell.

Every one of the feasible expectations within the solution polytope for the objective function would respect the bounds specified by any appropriate Bell-type inequality. It would not matter which coefficients on the product components of s were proposed as -1 and as $+1$ for the GHSZ problem. End of story.

Concluding technical comment

Quite contrary to the GHSZ conclusion that the premises of EPR pose a contradiction to a quantum experiment involving more than two particles, we can conclude that such experiments indeed do allow the premises of EPR, appropriately identified. Furthermore, the conditions of such an experiment might easily exemplify well the EPR premise of "Perfect correlation", if desired, in both the case of conditions of (11^*a) and that of (11^*b) which had confused GHSZ ... though not at the same time! Smile. This could be achieved by choosing the eight magnet angles involved as

$$(\phi_1, \phi_2, \phi_3, \phi_4, \phi_1', \phi_2', \phi_3', \phi_4') = (0, 0, 0, 0, \pi, \pi, \pi, \pi).$$

The value of κ^* associated with any choice of none, one, two, three, or four of the multiplicand angles of $A(\phi_1^*)B(\phi_2^*)C(\phi_3^*)D(\phi_4^*)$ as primed angles ϕ_i' rather than merely as ϕ_i would be either $0, \pi, -\pi, 2\pi,$ or -2π in every case. Thus, the Expectation of spin-products $E[A(\phi_1^*)B(\phi_2^*)C(\phi_3^*)D(\phi_4^*)] = -cos(\phi_1^* + \phi_2^* - \phi_3^* - \phi_4^*) = -cos(\kappa^*$ equals either -1 or $+1$ in every case.

From our original analysis of the GHSZ scenarios, we are now well aware that the realm matrix of possibilities for any one of the four-products is either $\mathbf{R}_{-1}$ or $\mathbf{R}_{+1}$, displayed in our equations (i) and (ii). Thus, any mixture distribution over these realm columns would yield the desired expectation in every case.

The troublesome results that had puzzled GHSZ before their shenanigans have been completely resolved. There is no contradiction at all inherent in positing local realism and supplementary variables as an explanation of quantum theory in dimensions higher than two. We shall clarify the mathematical structure of such a proposition in the next chapter.

Concluding comments

The results of this discussion have laid bare the claims of GHSZ, who denigrate the logical consistency of the EPR version of "local realism". However, I would not like a reader to think that I accord completely with the EPR construal of the situation either. From a larger perspective, I find it embedded in a view of physical experience that is seriously out of date. My investigations during the past nine years have been oriented quite narrowly, to a resolution of the quantum conundrums posed by Bell's inequality in the several forms in which it has been promoted. At a deeper level, I conclude that the so-called mysterious properties of the quantum world, involving a structure of "quantum probabilities" that differs from that of the mundane probabilities pertinent to the classical scale, have been misconstrued as well. A discussion of larger implications relevant to a reconstruction of physical theory awaits a confirmation of my narrow concerns. We shall broach some of them in Chapter 6.

As to the definitive empirical work of Hensen et al. (2016), I do not doubt their experimental results. However, their statistical analysis suffers from the same mistake of neglect as does that of Aspect, which we reviewed in Chapter 1.

Acknowledgements

This chapter constitutes an improved version an article that was first published in an issue of *Entropy* 2020, **22**, 0759. The contents of the Appendix that follows, concerning the empirical work of Pan et al., appear in the main body of the text of the published version.

Appendix: on Empirical evidence of Pan et al.

The following terse comments are meant exclusively for any reader who is familiar with the setup, the notation, and the details of the respected article of Pan, Bouwmeester, Daniell, and Zeilinger (2000), which I shall neither illuminate nor review here. Their developments continue from a presentation in an article by Greenberger, Horne, and Zeilinger (1989), referred to as GHZ below. I presume that any reader is up with the play. The empirical programme of this group relies on two erroneous presumptions.

In the first place, the joint activation of three different polarisation designs $Y_1Y_2X_3$, $Y_1X_2Y_3$, and $X_1Y_2Y_3$ cannot be performed on the same triplet of photons. In deference to the general uncertainty principle, quantum theory explicitly avoids any claims regarding their simultaneous result. What is proposed is a "thought experiment" to which the implications of locality would pertain. This would assert, for example, that the value of Y_2 in the instantiation of a $Y_1Y_2X_3$ design must be identical to its value in the simultaneous activation of an $X_1Y_2Y_3$ design *on the same triple of photons*. This is a claim widely recognised as lying outside the domain of quantum theory, since the results of such a programme cannot possibly be observed.

Secondly, in the observation of results from three *different* triples of photons proposed by the Pan group, the conditions of quantum superposition assure only that the *product* values of $Y_1 Y_2 X_3$ and $X_1 Y_2 Y_3$ are both identically equal to -1. They do not require that any *individual multiplicands* of the two product triples are equal. In particular, the value of Y_2 in the first experimental triple need not equal the value of Y_2 in a second triple. The implication proposed in the published analysis regarding the value of $X_1 X_2 X_3$ in any experimental run on three different triples of photons meeting the GHZ condition simply does not hold. The authors' allusion to experimental error as accounting for their mixed results does not wash. I would be happy to be more explicit in discussion or in a detailed article, but I will leave these comments in this terse form for now.

Chapter 3

RESURRECTING LOCAL REALISM :
the prospect of supplementary variables

A myriad of commonly accepted tenets of theoretical quantum physics are replete with error. Deliberations of the past half century regarding Bell's inequality have led to a widely accepted proposition that a non-contradictory formulation of Einstein's supplementary variables interpretation of quantum theory is impossible. In his memorial review of the situation some twenty years after publishing his initial empirical results, Alain Aspect (2002) claimed to have established that quantum theory conflicts with any supplementary parameter theory, since it violates the inequality. The claim was based on a mistaken derivation of the expectation of a crucial quantity arising in CHSH form as $E(s) = 2\sqrt{2}$, a value outside of the bounding interval of $[-2, +2]$ required by Bell. The culprit in the violation was identified as the principle of local realism.

The response of the physics community has been largely an acclamation of the "mysteries of the quantum world", promoting the strange phenomena that operate within a structure of probabilities differing from one pertinent to events observed on a larger scale. Replete with their allusions to parallel worlds, physical phenomena that require observation to instantiate, and wave activities that collapse to particles when observed, promoters of quantum mechanics have relished the acclaimed role of this branch of science in recognising astonishing features of nanoreality. Despite some attempts to mitigate the consequences of Bell violations or to qualify the bite of their implications, the possibility of fundamental errors in their propagation has not been seriously entertained.

To the contrary, we have learned in Chapter 1 that the deliberations of quantum theory yield only an interval bound on the expectation $E(s)$ as $(1.1213, 2]$, completely within the required interval. This conclusion was reached without reference to the proposition of supplementary variables at all, relying only on the structure of the gedankenexperiment involved under the presumption of local realism. Our attention in this present chapter will focus specifically on a mathematical formalisation of the supplementary variables proposition. Yet it reaches an identical conclusion. Without reference to the extensive contestable literature on related foundational matters, I turn directly to the mathematical issues at hand, proceeding with a constructive analysis which I now introduce.

We shall begin by embellishing the structure that Aspect proposed, so as to complete its characterisation and analysis in terms of the proposition of supplementary variables. We shall find that the results support the assertion of *any* probability distribution whatsoever regarding the four paired outcomes of the experiment, QM probabilities included. This understanding contrasts with that of Aspect, who thought that quantum theoretical probabilities are not congruent with hidden variables, while another probability distribution he termed the "naive" distribution *is* congruent, and is mistaken. Further, our assessment will make evident that every coherent distribution whatsoever supports an expected value of the CHSH quantity lying within its required bounds.

There is no coherent joint distribution for all four polarisation gedankenproducts that admits the quantum theoretic probability specifications for real experiments as its marginal distributions. Of course, only one of the experiments can actually be conducted, and the quantum probabilities are surely appropriate to any one of them. This feature concurs with its early recognition by Arthur Fine (1982), though I believe he overstated his case in one respect, which will be identified. Once again, we shall formulate a linear programming structure that identifies bounds on the quantum expectation $E(s)$ in keeping with the supplementary variables proposal. It too yields only an interval for the expectation $E(s)$ as $(1.1213, 2]$, rather than the mistaken unique evaluation of $2\sqrt{2}$ that is commonly subscribed. The interval arising from explicit recognition of the supplementary variables proposal is the same interval we computed in Chapter 1 without reference to it. The chapter will conclude with a brief discussion of pertinent issues.

It is ironic that in Bell's (1971) paper challenging the axiom-based conclusions of von Neumann (1932, 1955) regarding the impossibility of QM formulations involving supplementary variables, his thoughts were very close to identifying the basis of the conundrum his work had introduced. For he isolated the axiom that he found disputable as a requirement for any theory of quantum mechanics: "that any linear combination of quantities specified by observable Hermitian operations identifies an observable quantity whose expectation must equal the same linear combination of its component quantities." An enlightening discussion appears in Freire (2015), who reproduced this quotation (page 243). Nonetheless, Bell's consideration of the relevance of this axiom to his inequality in the context of a gedankenexperiment on a single pair of photons was unfinished, and he lost the opportunity for understanding the situation in the way we shall provide here.

Of course, it is always true that the expectation of a linear combination of quantities equals the same linear combination of their expectations. However, when one of those quantities is a function value of the other three, then its expectation is a function of *their* joint distribution. Quantum theory eschews the specification of such a joint distribution, as Fine and others have subsequently recognised. Thus, the expectation analysis does not yield a precise number, but rather merely a bounding interval. This can be identified by another application of Bruno de Finetti's fundamental theorem of probability, as we shall see.

As has transpired, claims to the impossibility of hidden variables formulations have largely won the day (supported even by Bell in higher dimensions), constituting a formidable literature on the topic. See Mermin (1993) and Mermin and Schack (2018), for examples. Recognition now of the errors involved in proclamations of the defiance of Bell's inequality allows for a reconsideration of both Einstein's principle of local realism and his counterproposal of unknown supplementary variables as the basis for the probabilistic specifications of quantum theory. Mathematical formalities of such a proposal constitute the subject matter of the present chapter.

A proposed explanation
of quantum behaviour

Refreshing the setup we explained in Chapter 1, we shall again use Aspect's notation, and expand upon it. An experiment is conducted on a pair of photons travelling in opposite directions along an axis, $\mathbf{z}$, from a common source. The direction of one photon travelling toward observation station A is opposite to the direction its paired photon travels toward station B: $\mathbf{z}_A = -\mathbf{z}_B$. At the end of their respective journeys, the photon paths are recorded by detectors identifying whether each of them passes through or is deflected by a polariser. At station A its linear crystal structure is angled in the (x, y) plane perpendicular to the incoming photon either in direction $\mathbf{a}$ or $\mathbf{a}'$. At station B the polariser is aligned either in direction $\mathbf{b}$ or $\mathbf{b}'$. This setup yields a specific relative angle between the polariser directions at A and B, as viewed in a common coordinate system. Using notation that parentheses around a pair of directions denotes the angle between them, the determination of the angles $(\mathbf{a}, \mathbf{z}_A)$ and $(\mathbf{b}, \mathbf{z}_B)$ implies the relative angle between the polarisation directions at the two stations is $(\mathbf{a}, \mathbf{b})$.

We begin exactly as Aspect does (with his equation 17 from the 2002 presentation) by defining a quantum quantity as a function of a vector λ. The components of this vector are posited as numerical indicators of unknown physical conditions (supplementary, or "hidden" variables) of the optical experiment we are considering, but which are not designated in the formulations of quantum theory. The value of $s(\lambda)$ is defined by results of a gedankenexperiment on the polarisations of a single pair of photons in four different relative angle settings:

$$s(\lambda) \equiv A(\lambda, \mathbf{a}) B(\lambda, \mathbf{b}) - A(\lambda, \mathbf{a}) B(\lambda, \mathbf{b}')$$
$$+ A(\lambda, \mathbf{a}') B(\lambda, \mathbf{b}) + A(\lambda, \mathbf{a}') B(\lambda, \mathbf{b}'), \text{ for } \lambda \in \Lambda. \quad (1)$$

The space of all possible values of such hidden variable vectors is denoted by Λ, of whatever dimension might be appropriate. The component functions $A(\cdot, \cdot)$ and $B(\cdot, \cdot)$ have the form

$$A(\lambda, \mathbf{a}) = +1 \ (\lambda \in \Lambda_{\mathbf{a}+}) - 1 \ (\lambda \in \Lambda_{\mathbf{a}-}), \text{ and}$$
$$B(\lambda, \mathbf{b}) = +1 \ (\lambda \in \Lambda_{\mathbf{b}+}) - 1 \ (\lambda \in \Lambda_{\mathbf{b}-}), \quad (2)$$

where the subspace pairs $(\Lambda_{\mathbf{a}+}, \Lambda_{\mathbf{a}-})$ and $(\Lambda_{\mathbf{b}+}, \Lambda_{\mathbf{b}-})$ provide distinct partitions of Λ. They separate the possible settings of λ into those which might yield the polarisation observation at each station as -1 or $+1$.

Here and throughout this work, the use of parentheses surrounding a mathematical statement that may be true or false denotes the indicator value of 1 if the statement is true, and 0 if it is false. This convention provides that the values of $A(\lambda, \mathbf{a})$ and $B(\lambda, \mathbf{b})$ defined in (2) might each equal only either -1 or $+1$. For example, λ must be an element of either $\Lambda_{\mathbf{a}+}$ or $\Lambda_{\mathbf{a}-}$, but not both. The two partitions of Λ are understood to be distinct, so their Cartesian product identifies a 4-constituent partition of

$$\Lambda \;=\; \Lambda_{\mathbf{a}+}\Lambda_{\mathbf{b}+} \;\cup\; \Lambda_{\mathbf{a}+}\Lambda_{\mathbf{b}-} \;\cup\; \Lambda_{\mathbf{a}-}\Lambda_{\mathbf{b}+} \;\cup\; \Lambda_{\mathbf{a}-}\Lambda_{\mathbf{b}-} \,. \qquad (3)$$

Before continuing our analysis of the situation, we should note aloud that the principle of local realism would support a factorised representation of $s(\lambda)$ in equation (1) as either

$$A(\lambda, \mathbf{a})\,[B(\lambda, \mathbf{b}) - B(\lambda, \mathbf{b}')] \;+\; A(\lambda, \mathbf{a}')\,[B(\lambda, \mathbf{b}) + B(\lambda, \mathbf{b}')]$$

or as $\quad B(\lambda, \mathbf{b})\,[A(\lambda, \mathbf{a}) - A(\lambda, \mathbf{a}')] \;+\; B(\lambda, \mathbf{b}')\,[A(\lambda, \mathbf{a}) + A(\lambda, \mathbf{a}')]\,.$

For the principle provides that when an observation of $A(\lambda, \mathbf{a})$ is made in consort with an observation $B(\lambda, \mathbf{b})$, as required to evaluate empirically the first summand of $s(\lambda)$, whatever value A takes in this experiment is understood to be the same numerical value it would take on when observed in consort with $B(\lambda, \mathbf{b}')$ instead. For the value of A is presumably guided by the setting indicated by λ. The resulting observation indicator appears as the second summand of $s(\lambda)$. Although the quantum probabilities for the observation values of A and B do depend on the relative angles $(\mathbf{a}, \mathbf{b})$ and $(\mathbf{a}, \mathbf{b}')$ in two such experimental situations, it is the presumed principle of local realism that would allow us to factor the individual observation value of $A(\lambda, \mathbf{a})$ out of these two terms. We had detailed in Chapter 1 the reasoning that implies the value of $s(\lambda)$ is thus restricted to equal only either -2 or $+2$. Enough said.

The idea behind this formalisation of the observation functions $A(\lambda, \mathbf{a})$ and $B(\lambda, \mathbf{b})$ is that the summary information we have explained regarding the polariser directions is deemed insufficient to describe the situation of the quantum experiment completely. For substantive examples, we have left unmentioned the unknown phase position of the photon wave motions when they reach an interactive distance from the polarising materials at stations A and B. This means we do not know precisely the angle of the transverse direction at which either photon interacts with its polariser, moving as a wave. So we do not know the transverse velocity of the particle at detection interaction either, not even to mention the possible relevance of its unobserved quark structure. Nor do we know the actual specific molecular crystal structure of the material surface at the exact region where the photon is to be engaged. This too surely involves some variability. Indeed, with our current state of technology it is hard to imagine how we might ever be able to observe such features of the experimental "happening" during the physical transaction when the polarising activities are being recorded. Moreover, there may well be other natural features of the happening relevant to the occurring polarisation activity that we cannot currently imagine at all. However, it is proposed that if we knew enough of such "hidden" details of the experimental operation, we would be able to know for certain what the spin observations would be. That is what the hidden variable dimensions of the incompleteness argument are all about.

Such a scenario would characterise the mechanics of quantum activity in a fashion similar to that of physical activity at the scale of everyday life. The laws of force regarding the mechanics of motion at this scale are fairly definitive. The reason we cannot predict the result of a ten-pin bowl is merely that we do not know precisely the angular direction at which and acceleration with which the bowling ball leaves the bowler's hand, nor the spin angle, nor the spin velocity of the ball, nor the coefficient of friction of the lanes in their current preparation. At least we can specify what the relevant factors would be. With such knowledge, the laws of motion would specify the result of the bowl. At the quantum scale, however, the theory of physical mechanics provides us not an exact specification of experimental observations as a function of the variables we know, but merely probability specifications for results we might observe. Einstein claimed that the structure of the situations at quantum and classical scales are actually the same. It is just that at the quantum scale we cannot even specify what all the relevant conditioning variables are, much less their precise values during the conduct of any experiment. John Bell originally suspected the same thing, though he was bemused by his inequality and puzzled by its apparent violation.

Despite its relegation in mainstream quantum literature, Ed Jaynes (1986) was an eminent advocate of the proposition that supplementary variables should be relevant to the characterisation of quantum phenomena, for reasons he explained well, and he supported their investigation. He has not been alone. By now there are a number of fronts on which such research has been engaged, although results are controversial and sometimes merely speculative. For examples, these include widely broadcast efforts on the pilot wave front (see Falk (2016) for some popular overview), a wide variety of investigations referenced on the Orcid site of Bassi (2023), and ambitious mathematical detail of Santilli (1996) concerning the characterisation of deformable particles.

Of course, such theoretical formalisation and experimentation are not easy, as hidden variables have remained hidden. It would seem that such investigations should be welcomed and promoted, rather than being dismissed as impossible. It is hard to imagine that specification of the relative angle between polarisers or between the magnet directions in a paired Stern-Gerlach experiment would exhaust all that could possibly be known about photon behaviour or electron spin experiments. Such a simplification of mysterious quantum activity is hardly merited by the mistaken defiance of Bell's inequality by quantum probabilities in a gedankenexperiment. A serious literature review of the current status of such research is beyond the scope of this book. My aim here is merely to formalise a mathematical structure for the proposal of supplementary variables that cannot be dismissed as impossible. The coherent bounds for the CHSH expectation $E(s)$ will emerge as a by-product, agreeing with the results appearing in Lad (2021).

The supplementary variables partition

Numerical summary measures of proposed supplementary variables are meant to be represented by components of a vector, λ. For a polariser set up in direction $\mathbf{a}$, for example, the function $A(\lambda, \mathbf{a})$ is meant to represent the detection recorded at station A under hidden conditions that would be recorded as λ, if they were measured. The domain of all *possible valuations* of such λ vector variables is denoted by Λ. It can be separated into two exclusive and exhaustive pieces, denoted by $\Lambda_{\mathbf{a}+}$ and $\Lambda_{\mathbf{a}-}$, these being the collections of such possibilities that would give rise to a measurement of $A(\lambda, \mathbf{a}) = +1$ and to $A(\lambda, \mathbf{a}) = -1$, respectively.

This form of the component function pairs, $[A(\lambda, \mathbf{a}), B(\lambda, \mathbf{b})]$, also pertains to pairs in which either or both of the spin detector directions ($\mathbf{a}$ and $\mathbf{b}$) are replaced by $\mathbf{a}'$ and $\mathbf{b}'$, in exactly the same way. In accounting for them all, we would have four such paired functions defined and observed, corresponding to the polariser angles $(\mathbf{a}, \mathbf{b}), (\mathbf{a}, \mathbf{b}'), (\mathbf{a}', \mathbf{b})$, and $(\mathbf{a}', \mathbf{b}')$ that are entertained in Bell's gedankenexperiment. The Cartesian product of the four such Λ partitions then yield a refined 16-constituent partition of Λ, viz.,

$$
\begin{aligned}
\Lambda = \; & \Lambda_{a+}\Lambda_{b+}\Lambda_{a'+}\Lambda_{b'+} \cup \Lambda_{a+}\Lambda_{b+}\Lambda_{a'+}\Lambda_{b'-} \cup \Lambda_{a+}\Lambda_{b+}\Lambda_{a'-}\Lambda_{b'+} \cup \Lambda_{a+}\Lambda_{b+}\Lambda_{a'-}\Lambda_{b'-} \cup \\
& \Lambda_{a-}\Lambda_{b+}\Lambda_{a'+}\Lambda_{b'+} \cup \Lambda_{a-}\Lambda_{b+}\Lambda_{a'+}\Lambda_{b'-} \cup \Lambda_{a-}\Lambda_{b+}\Lambda_{a'-}\Lambda_{b'+} \cup \Lambda_{a-}\Lambda_{b+}\Lambda_{a'-}\Lambda_{b'-} \cup \\
& \Lambda_{a+}\Lambda_{b-}\Lambda_{a'+}\Lambda_{b'+} \cup \Lambda_{a+}\Lambda_{b-}\Lambda_{a'+}\Lambda_{b'-} \cup \Lambda_{a+}\Lambda_{b-}\Lambda_{a'-}\Lambda_{b'+} \cup \Lambda_{a+}\Lambda_{b-}\Lambda_{a'-}\Lambda_{b'-} \cup \\
& \Lambda_{a-}\Lambda_{b-}\Lambda_{a'+}\Lambda_{b'+} \cup \Lambda_{a-}\Lambda_{b-}\Lambda_{a'+}\Lambda_{b'-} \cup \Lambda_{a-}\Lambda_{b-}\Lambda_{a'-}\Lambda_{b'+} \cup \Lambda_{a-}\Lambda_{b-}\Lambda_{a'-}\Lambda_{b'-} \, .
\end{aligned}
\tag{4}
$$

It is worth noticing here for future reference that the union of only the first four row-constituents of this partition equals the first constituent of the simpler partition displayed in (3). The unions of each of the next three rows would identify the other constituents of that partition. Moreover, unions of other carefully chosen components of (4) would provide us with three other partitions of Λ relevant to the component experiments of the gedankenexperiment, as we shall find to be useful. For an example, taking unions down the columns of the display would identify another easy partition:

$$
\Lambda = \Lambda_{a'+}\Lambda_{b'+} \cup \Lambda_{a'+}\Lambda_{b'-} \cup \Lambda_{a'-}\Lambda_{b'+} \cup \Lambda_{a'-}\Lambda_{b'-} \, .
$$

We shall need to poke around to find the two others. For later reference in this endeavour, let us enumerate the partition components of equation (4) as Λ_1 through Λ_{16}, numbering them sequentially across the rows as they appear in this partition equation.

Considering both possible directions for the polarisers at A and B (these being $\mathbf{a}$ or $\mathbf{a}'$, and $\mathbf{b}$ or $\mathbf{b}'$, respectively) along with the possible spin measurements of $+1$ or -1 at each end of such a pairing, the domain Λ is thus partitioned into the 16 constituents whose members are listed in the partition equation (4). According to the imagination of the hidden variables proposition, the conceivably observable but hidden value of the λ vector would be found to be within one of these sixteen constituents of its domain partition. Whichever one it happens to be, the value of $s(\lambda)$ would be observed. For examples, evaluating the summands of Bell's quantity according to the functions specified, we find

if $\lambda \in \Lambda_{a-}\Lambda_{b-}\Lambda_{a'+}\Lambda_{b'+}$,

then $s(\lambda) = (-1)(-1) - (-1)(+1) + (+1)(-1) + (+1)(+1) = 2,$

or if $\lambda \in \Lambda_{\mathbf{a}+} \Lambda_{\mathbf{b}-} \Lambda_{\mathbf{a}'+} \Lambda_{\mathbf{b}'+}$,

then $s(\lambda) = (+1)(-1) - (+1)(+1) + (+1)(-1) + (+1)(+1) = -2$,

or if $\lambda \in \Lambda_{\mathbf{a}-} \Lambda_{\mathbf{b}+} \Lambda_{\mathbf{a}'-} \Lambda_{\mathbf{b}'-}$,

then $s(\lambda) = (-1)(+1) - (-1)(-1) + (-1)(+1) + (-1)(-1) = -2$.

Evaluating the value of $s(\lambda)$ for the λ values in every one of the constituents of the partition of Λ would show that the only possible values for s are -2 and $+2$, identifying the realm of possibilities $\mathcal{R}(s) = \{-2, +2\}$.

Now it is clear that *whatever* probabilities might be associated with the constituents of the partition of Λ, the expected value of the quantity s defined in (1) would yield

$$E[s(\lambda)] = E[A(\lambda, \mathbf{a}) B(\lambda, \mathbf{b})] - E[A(\lambda, \mathbf{a}) B(\lambda, \mathbf{b}')]$$
$$+ E[A(\lambda, \mathbf{a}') B(\lambda, \mathbf{b})] + E[A(\lambda, \mathbf{a}') B(\lambda, \mathbf{b}')]. \quad (5)$$

Expectation is understood to be evaluated here with respect to some distribution over $\lambda \in \Lambda$ which admits a density $\rho(\lambda)$. Evaluated over the constituents of the 16-component partition of Λ, it would generate a probability mass function: that is, a schedule of probabilities for the components of the partition of Λ that derive from this density. This equation (5) mimics that of Aspect's equation numbered (21), which he labels as his quantity S. This expectation must lie within $[-2, +2]$, the convex hull of the realm of possibility for the linear combination of polarisation products itself, which is labeled s. This specifies Bell's inequality in this context: $-2 \leq E(s) \leq +2$.

Arthur Fine (1982) had already recognised the reparametrization of the expectation $E(s)$ that could be afforded by an appropriate partition of the space Λ. It was his understanding that such a characterisation of the problem *requires* the specification of a *complete* distribution over the results of the four incompatible observations. We shall find that such a complete distribution is not required to determine some implications of quantum theory for the value of $E(s)$. This mistaken idea distracted quantum theoretical considerations from the usefulness of the partition. For quantum theory evades such a complete specification. We shall see now that quantum theoretic specifications regarding the outcomes of the four possible experiments that *can* be conducted are sufficient to specify a precise interval bound on $E(s)$. This merely reduces the space of cohering distributions to a convex polytopic subspace of the full space of distributions, spanned by the possible outcomes of the gedankenexperiment. We shall observe the relevance of this remark in a discussion that will conclude this chapter.

However, we should first examine expectation (5) in more detail, studying specifically its first summand expectation, which will be paradigmatic for each of the four. Evaluating expectation with respect to any quantum theoretic mass function appropriate to the partition $\{\Lambda_{\mathbf{a}+}\Lambda_{\mathbf{b}+}, \Lambda_{\mathbf{a}+}\Lambda_{\mathbf{b}-}, \Lambda_{\mathbf{a}-}\Lambda_{\mathbf{b}+}, \Lambda_{\mathbf{a}-}\Lambda_{\mathbf{b}-}\}$ would yield

$$
\begin{aligned}
E[A(\lambda, \mathbf{a})B(\lambda, \mathbf{b})] &= (+1)P[\Lambda_{\mathbf{a}+}\Lambda_{\mathbf{b}+}] + (-1)P[\Lambda_{\mathbf{a}+}\Lambda_{\mathbf{b}-}] \\
&\quad + (-1)P[\Lambda_{\mathbf{a}-}\Lambda_{\mathbf{b}+}] + (+1)P[\Lambda_{\mathbf{a}-}\Lambda_{\mathbf{b}-}] \\
&= P[A(a)B(b) = +1] - P[A(a)B(b) = -1] \\
&= 2\,P[A(a)B(b) = +1] - 1 \\
&= 2\{P[\Lambda_{\mathbf{a}+}\Lambda_{\mathbf{b}+}] + P[\Lambda_{\mathbf{a}-}\Lambda_{\mathbf{b}-}]\} - 1. \quad (6)
\end{aligned}
$$

That second equality holds because quantum probabilities respect the equalities $P_{++} = P_{--}$ and $P_{-+} = P_{+-}$. Now the same form of these four lines of evaluation would apply if either or both of the polariser directions $\mathbf{a}$ and $\mathbf{b}$ were replaced therein by $\mathbf{a}'$ or $\mathbf{b}'$, respectively. Recognising that these developments rely on the supposition of local realism, as described above, we have completed our construction of the hidden variables setup. We are ready for an analysis of $E[s(\lambda)]$ with respect to any distribution in the cohering QM polytope over the sixteen constituents that partition Λ.

We conclude this introduction to the hidden variables interpretation of quantum expectations by noting that the representations of equation (6) do not preclude *any* coherent expectations of the products of spins at A and B whatsoever, just so long as they respect the symmetry conditions involving $P_{++} = P_{--}$ and $P_{-+} = P_{+-}$. It would be useful for representing both QM-motivated probabilities as well as the expectations Aspect refers to as the "naive supplementary model", for example. Both of these structures of spin expectations can be represented by hidden variables parametrisations. Because this understanding conflicts with that of Aspect, who thought that QM probabilities cannot be parameterised by hidden variables while his proposed "naive model" can, we should defer our analysis for a while to dwell briefly on this thought.

Substantive content of an HV parametrization

Hidden variables motivation for an assessment of uncertainty about any quantity are merely considerations that identify a reparametrization of the quantity in question. They may be helpful in someone's assessment of the expectation, and they may not. They are surely not *required* for the assertion of relevant probabilities. Identical probability assertions regarding the observable quantities might be promoted both by someone who thinks about an experimental situation in terms of hidden variables and by someone who does not. This is the content of the equivalence of the lines of equation (6). The two viewpoints are observationally equivalent, for the hidden variables are, of course, hidden.

Mathematically, hidden variables theory amounts merely to a one-to-one transformation of the partitioned space of possibilities for the four quantities composing s to the 16-constituent partition of hidden variables space. The space of observation possibilities for
$$[\, A(\lambda, \mathbf{a})B(\lambda, \mathbf{b}),\ A(\lambda, \mathbf{a})B(\lambda, \mathbf{b}'),\ A(\lambda, \mathbf{a}')B(\lambda, \mathbf{b}),\ A(\lambda, \mathbf{a}')B(\lambda, \mathbf{b}')\,]$$
is mapped onto the partitioned space of Λ as defined in equation (4). There is nothing impossible about it at all.

The former partition of the possible observation values contains the sixteen (4×1) vectors whose components each equal either -1 or $+1$; and each of these vectors contains either none, one, two, three or four -1's, in any order. The latter partition is a partition of the hidden variables space, Λ. The probabilities one might assess for the observation vector partition must be the same as those one might assess for the hidden variables. Of course, one cannot directly assess the probabilities for the hidden variables, because they are hidden. Einstein's proposition merely imagined them as explanations of his viewpoint that the theory of quantum mechanics must be incomplete. The probabilities he would assert for the observable spin values, while in ignorance of such hidden variables and their values, are nonetheless identical to the probabilities of quantum theorists who imagine that the QM probabilities are actually inherent in the photons. Basically, the assertion values derive from the symmetry of our uncertain opinions about the situations of the two photons as they engage their polarisers. We will illuminate this insight in Chapter 6.

As a bottom line, a hidden variables explanation of polarisation product expectations can apply to any probability distribution for the possible observation values whatsoever, including those moti-vated by the theory of quantum mechanics. Let's get down to the business of assessing the general form of an expectation for the Aspect/CHSH/Bell quantity, s, and particularly the expectation motivated by quantum theory.

Assessing $E(s)$ via entangled distributions

We shall now address the assessment of $E(s)$, and in particular a surprising identification of the quantum expectation that challenges the Aspect/Bell mistaken assertion that $E_{QM}(s) = 2\sqrt{2}$ still again. Recall that

$$E[s(\lambda)] = E[A(\lambda, \mathbf{a})\, B(\lambda, \mathbf{b})] - E[A(\lambda, \mathbf{a})\, B(\lambda, \mathbf{b}')]$$
$$+ E[A(\lambda, \mathbf{a}')\, B(\lambda, \mathbf{b})] + E[A(\lambda, \mathbf{a}')\, B(\lambda, \mathbf{b}')].$$

Using the *form* of equation (6) for each of these four expected prod-uct summands, which would be identical for each, viz.,

$$E[A(\lambda, \mathbf{a})B(\lambda, \mathbf{b})] = 2\{P[\Lambda_{\mathbf{a}+}\Lambda_{\mathbf{b}+}] + P[\Lambda_{\mathbf{a}-}\Lambda_{\mathbf{b}-}]\} - 1,$$

and performing the summations appropriate to the definition of s (understood to involve one negative sign), we can then write

$$\begin{aligned} E[s(\lambda)] = 2\{\, & P[\Lambda_{\mathbf{a}+}\Lambda_{\mathbf{b}+}] + P[\Lambda_{\mathbf{a}-}\Lambda_{\mathbf{b}-}] \\ & + P[\Lambda_{\mathbf{a}'+}\Lambda_{\mathbf{b}+}] + P[\Lambda_{\mathbf{a}'-}\Lambda_{\mathbf{b}-}] \\ & + P[\Lambda_{\mathbf{a}'+}\Lambda_{\mathbf{b}'+}] + P[\Lambda_{\mathbf{a}'-}\Lambda_{\mathbf{b}'-}] \\ & - P[\Lambda_{\mathbf{a}+}\Lambda_{\mathbf{b}'+}] - P[\Lambda_{\mathbf{a}-}\Lambda_{\mathbf{b}'-}]\} - 2. \end{aligned} \quad (7)$$

Now, our next insight will look quite messy, algebraically, but what we need to do is to register the fact that probabilities for these various partition constituents are necessarily related to one another. Each of them arises from assessing uncertainty regarding the sum of four constituents of the 16-constituent partition of Λ we detailed in equation (4). For each of the probability summands appearing in equation (7), we need to identify which components of the 16-component partition of Λ require a probability assessment. For example, the first partition constituent, $\Lambda_{\mathbf{a}+}\Lambda_{\mathbf{b}+}$, which is as-sessed with the first probability appearing in (7), is composed of the union of four constituents of the finer partition of Λ, viz.,

$$\Lambda_{\mathbf{a}+}\Lambda_{\mathbf{b}+} = \Lambda_{\mathbf{a}+}\Lambda_{\mathbf{b}+}\Lambda_{\mathbf{a}'+}\Lambda_{\mathbf{b}'+} \cup \Lambda_{\mathbf{a}+}\Lambda_{\mathbf{b}+}\Lambda_{\mathbf{a}'-}\Lambda_{\mathbf{b}'+}$$
$$\cup \Lambda_{\mathbf{a}+}\Lambda_{\mathbf{b}+}\Lambda_{\mathbf{a}'+}\Lambda_{\mathbf{b}'-} \cup \Lambda_{\mathbf{a}+}\Lambda_{\mathbf{b}+}\Lambda_{\mathbf{a}'-}\Lambda_{\mathbf{b}'-}.$$

These are the constituents we agreed to number $1, 2, 3$ and 4. Thus, $P[\Lambda_{a+}\Lambda_{b+}]$ must be accorded with the sum of their probabilities. In order to clinch the messy algebraic implications of this recognition, I present the complete composition of $E[s(\lambda)]$ below, using a schematic format. The three blocks of terms headed with a "+" in their top left-hand corner correspond to the pairs of summands in the first three rows of the display of (7), and the final block headed with a "−" corresponds to the final row, which involves a subtraction.

A Schema Enumerating Constituents of the Partition of Λ

+

*1	$\Lambda_{a+}\Lambda_{b+}$	$\Lambda_{a'+}\Lambda_{b'+}$	+	$\Lambda_{a-}\Lambda_{b-}$	$\Lambda_{a'+}\Lambda_{b'+}$	*13
*2		$\Lambda_{a'+}\Lambda_{b'-}$	+		$\Lambda_{a'+}\Lambda_{b'-}$	*14
*3		$\Lambda_{a'-}\Lambda_{b'+}$	+		$\Lambda_{a'-}\Lambda_{b'+}$	*15
*4		$\Lambda_{a'-}\Lambda_{b'-}$	+		$\Lambda_{a'-}\Lambda_{b'-}$	*16

+

*1	$\Lambda_{a'+}\Lambda_{b+}$	$\Lambda_{a+}\Lambda_{b'+}$	+	$\Lambda_{a'-}\Lambda_{b-}$	$\Lambda_{a+}\Lambda_{b'+}$	*11
*2		$\Lambda_{a+}\Lambda_{b'-}$	+		$\Lambda_{a+}\Lambda_{b'-}$	*12
*5		$\Lambda_{a-}\Lambda_{b'+}$	+		$\Lambda_{a-}\Lambda_{b'+}$	*15
*6		$\Lambda_{a-}\Lambda_{b'-}$	+		$\Lambda_{a-}\Lambda_{b'-}$	*16

+

*1	$\Lambda_{a'+}\Lambda_{b'+}$	$\Lambda_{a+}\Lambda_{b+}$	+	$\Lambda_{a'-}\Lambda_{b'-}$	$\Lambda_{a+}\Lambda_{b+}$	*4
*9		$\Lambda_{a+}\Lambda_{b-}$	+		$\Lambda_{a+}\Lambda_{b-}$	*12
*5		$\Lambda_{a-}\Lambda_{b+}$	+		$\Lambda_{a-}\Lambda_{b+}$	*8
*13		$\Lambda_{a-}\Lambda_{b-}$	+		$\Lambda_{a-}\Lambda_{b-}$	*16

−

*1	$\Lambda_{a+}\Lambda_{b'+}$	$\Lambda_{a'+}\Lambda_{b+}$	+	$\Lambda_{a-}\Lambda_{b'-}$	$\Lambda_{a'+}\Lambda_{b+}$	*6
*9		$\Lambda_{a'+}\Lambda_{b-}$	+		$\Lambda_{a'+}\Lambda_{b-}$	*14
*3		$\Lambda_{a'-}\Lambda_{b+}$	+		$\Lambda_{a'-}\Lambda_{b+}$	*8
*11		$\Lambda_{a'-}\Lambda_{b-}$	+		$\Lambda_{a'-}\Lambda_{b-}$	*16

The expectation equation (7) for $E(s)$ says to sum firstly the probabilities for the three blocks of partition constituents headed by plus signs, and then to subtract the probability for the fourth constituent block headed by a minus. Then double this result and subtract 2. On the right-hand and left-hand sides of each of the four blocks displayed above appears an exhaustive list of constituent numbers from the 16-partition of Λ whose probabilities are to be summed (the first three blocks) or subtracted (the last block). Each constituent in the summable list is numbered. As can be seen there, each of the probabilities for constituents listed in the bottom block (to be subtracted) will cancel the probability for a constituent matching it in one of the first three blocks. Then the constituents remaining whose probabilities are to be summed are those numbered $1, 2, 4, 5, 12, 13, 15$ and 16; and each of these appears twice. Equation (7) says that to compute $E[s(\lambda)]$, this doubled sum should be then itself be doubled, and finally have the number 2 subtracted. Because the sum of the probabilities (whatever values they might have) for the numbered constituents remaining is surely within $[0, 1]$, its double is surely within $[0, 2]$. Doubling that number will yield a number within $[0, 4]$, and subtracting 2 according to the directions of equation (7), *will surely yield a number within the interval* $[-2, +2]$, just as required by coherency, and just as required by Bell's inequality.

This numerical analysis makes more convincing than ever my claim that the Aspect/Bell derivation of $E_{QM}(s) = 2\sqrt{2}$ is incorrect. We can conclude that Bell's inequality is not defied at all. It matters not what might be the probabilities tendered regarding the components of s, whether based on quantum theory, or on Aspect's caricature of naive realism, or whatever. Moreover, this result derives directly from the supposition of supplementary variables! Rather than being incongruous with the quantum theoretic assessment of $E(s)$, the proposition of hidden variables underlines the correct numerical assessment of this expectation as an interval, a programme we shall now engage.

A bounded computation of $E_{QMHV}(s)$

Codification of possible results of an Aspect/Bell gedankenexperiment in parametric form, as afforded by the partition of Λ in our developments of equations $(1, 2, 3)$, was recognised long ago in a notable (and contested) article by Arthur Fine (1982). The proposal of additional variables supplementary to the mere specification of the relative angles between the polarisers at stations A and B allows one to express quantum probabilities for the polarisation observations in terms of probabilities for components of the partition of Λ. With this I surely agree. However, I aver that Fine overstated the implication of his analysis when claiming (1982, p 291) that "the existence of a deterministic hidden-variables model is strictly equivalent to the existence of a joint probability distribution function $P(AA'BB')$ (sic) for the four observables of the experiment, one that returns the probabilities of the experiment as marginals."

Concluding my analysis of the situation in this Section, I shall now develop the bounding implications of the agreeable and powerful probabilistic assertions of quantum mechanics. These identify only a convex space of distributions cohering with its propositions, rather than a unique complete distribution as Fine suggested. Quantum theory does not provide for any such complete joint distribution for the incompatible observations of the gedankenexperiment, nor for the widely touted incorrect expectation specification of $E_{QM}(s) = 2\sqrt{2}$.

To begin, the formulation of the sixteen-component partition of the hidden variables space Λ in equation (3) does not presuppose the assertion of a probability distribution over these components. The analysis subsequent to this formulation has determined that the quantum theoretic expectation $E(s)$ would require merely the specification of probabilities for components $1, 2, 4, 5, 13, 15$, and 16 of this partition. Even at that, each of these probabilities would require an assertion regarding the joint outcome of four incommensurable experiments, and quantum theory explicitly avoids any such assertions. Where does this leave us?

Quantum theory is very explicit, nonetheless, in specifying a precise expectation value for the polarisation product of observations (and a probability distribution for the four possible paired observation results) at *any one* of the paired angles at which an experiment might actually be conducted, $(\mathbf{a}, \mathbf{b}), (\mathbf{a}, \mathbf{b}'), (\mathbf{a}', \mathbf{b})$, and $(\mathbf{a}', \mathbf{b}')$. In the context of Aspect's experimental design, the relative angle settings of the polariser directions at A and B are $-\pi/8, -3\pi/8, \pi/8$, and $-\pi/8$, respectively. These settings motivate the expectation values

$$E[A(\mathbf{a})B(\mathbf{b})] = E[A(\mathbf{a}')B(\mathbf{b})] = E[A(\mathbf{a}')B(\mathbf{b}')] = 1/\sqrt{2},$$
$$\text{and } E[A(\mathbf{a})B(\mathbf{b}')] = -1/\sqrt{2}.$$

These details are well known from the exposition of Aspect (2002) and have been described in Chapter 1. I will now rely on your examination of the partition equation (4) to recognise that the supplementary variables partition would correspond to a companion partition of the possible gedanken polarisation product observation vectors as shown in the columns of the following realm matrix:

$$
\mathbf{R}\begin{pmatrix} \Lambda_i \\ \ast\ast\ast\ast\ast \\ A(\mathbf{a}) \\ B(\mathbf{b}) \\ A(\mathbf{a'}) \\ B(\mathbf{b'}) \\ \ast\ast\ast\ast\ast \\ A(\mathbf{a})B(\mathbf{b}) \\ A(\mathbf{a})B(\mathbf{b'}) \\ A(\mathbf{a'})B(\mathbf{b}) \\ A(\mathbf{a'})B(\mathbf{b'}) \end{pmatrix} =
$$

Λ_1	Λ_2	Λ_3	Λ_4	Λ_5	Λ_6	Λ_7	Λ_8	Λ_9	Λ_{10}	Λ_{11}	Λ_{12}	Λ_{13}	Λ_{14}	Λ_{15}	Λ_{16}
1	1	1	1	-1	-1	-1	-1	1	1	1	1	-1	-1	-1	-1
1	1	1	1	1	1	1	1	-1	-1	-1	-1	-1	-1	-1	-1
1	1	-1	-1	1	1	-1	-1	1	1	-1	-1	1	1	-1	-1
1	-1	1	-1	1	-1	1	-1	1	-1	1	-1	1	-1	1	-1
1	1	1	1	-1	-1	-1	-1	-1	-1	-1	-1	1	1	1	1
1	-1	1	-1	-1	1	-1	1	1	-1	1	-1	-1	1	-1	1
1	1	-1	-1	1	1	-1	-1	-1	-1	1	1	1	1	-1	-1
1	-1	-1	1	1	-1	-1	1	-1	1	1	-1	1	1	-1	1

The first four numerical rows of this matrix identify the possible polarisation observations of the four components of the gedanken-experiment, and the final four rows represent the products of one of the A's with one of the B's. These final four rows contain only eight distinct columns, since the final eight columns of polarisation products repeat the first eight in reverse order. This alerts us to the fact that the value of any one of the those rows is related functionally to the other three via a mapping $\{-1, +1\}^3 \to \{-1, +1\}$. For example, the first eight columns of the first three rows exhaust the possibilities of $\{-1, +1\}^3$. In fact, this is true of any three rows of product possibilities chosen from the four. The remaining row is restricted to equal a value specified by these three. For use in what follows, let's name these four row vectors of polarisation product values as $\mathbf{r_{ab}}, \mathbf{r_{ab'}}, \mathbf{r_{a'b}}$, and $\mathbf{r_{a'b'}}$.

Of course, quantum theory does not specify a vector of probabilities, $\mathbf{q}_{16}$, for the column vectors of possible gedanken polarisation outcomes. It would only require that these vector components all be non-negative, and that they sum to 1. For the columns provide an exhaustive list of exclusive possibilities for the results of the gedankenexperiment. Nonetheless, quantum theory does say something specific about particular *linear combinations* of these unspecified probabilities. Each expectation of a polarisation product at a single angle setting specifies the value of a linear combination of them, viz.,

$$E[A(\mathbf{a})B(\mathbf{b})] = \mathbf{r_{ab}q}_{16}, \qquad E[A(\mathbf{a})B(\mathbf{b'})] = \mathbf{r_{ab'}q}_{16},$$

$$E[A(\mathbf{a'})B(\mathbf{b})] = \mathbf{r_{a'b}q}_{16}, \quad \text{and } E[A(\mathbf{a'})B(\mathbf{b'})] = \mathbf{r_{a'b'}q}_{16}.$$

However, the theory says nothing about products of more than two of the polarisation results, which cannot be observed. Moreover, the functional relations among the four components of the CHSH/Bell quantity s imply that only three of these displayed expectations can be recognised at a time in conditions under which Bell's inequality pertains ... when the gedankenexperiment is meant to be applied to the same two photons at all four of the angled polarisation settings. But these might be *any* three of them, and there are four choices of three that can be entertained.

The upshot of these considerations is that the implications of quantum theory can be identified by the specification of four paired linear programming problems. I shall first specify the LP problem algebraically, and then describe it. Here is one of the LP pairs:

Find the minimum and maximum values of

$$2 \left\{ [\, 2 \, (1\,1\,0\,1\,1\,0\,0\,0\,0\,0\,0\,1\,1\,0\,1\,1) \; \mathbf{q}_{16}] \right\} \, -2$$

subject to the linear restrictions

$$\begin{pmatrix} \mathbf{r_{ab}} \\ \mathbf{r_{ab'}} \\ \mathbf{r_{a'b}} \end{pmatrix} \mathbf{q}_{16} = \begin{pmatrix} 1/\sqrt{2} \\ -1/\sqrt{2} \\ 1/\sqrt{2} \end{pmatrix},$$

as required of the expectations that we have presumed specified, and where the components of $\mathbf{q}_{16}$ are non-negative, and sum to 1.

The other three pairs of min/max LP routines would derive from changing the choice of the constraining $\mathbf{r_{a^*b^*}}$ vectors to each of the other three possibilities.

The objective function vector $(1\,1\,0\,1\,1\,0\,0\,0\,0\,0\,0\,1\,1\,0\,1\,1)$ identifies the components of $\mathbf{q}_{16}$ involved in the representation of $E(s)$ as the sum of component numbers 1, 2, 4, 5, 12, 13, 15, and 16, as we noted in the discussion following the schematic formulation above. Appropriately, the objective function says to double this sum, and then double it again, and finally to subtract 2. The three linear constraining equations represent the restrictions required by the quantum theoretic specification of expectations for the polarisation products in three of the angle settings. We need not be more expansive in the description of this formulation here, because we have already considered ornate details of this type of computational argument in Chapter 1 while exposing the error involved in the Aspect/Bell problem.

While the detailed algebraic format of the computations I have described here are different from those appearing in the Aspect/Bell analysis, the problem the two formats resolve is precisely the same one. It is not surprising that the numerical solutions to the two sets of eight linear programming problems are identical. Again, duplications in the realm matrix reduce the dimension of $\mathbf{q}_{16}$ to 8. This is despite the fact that the formulation here relies explicitly on a characterisation of hidden variables, while the one portrayed in Chapter 1 does not. Quantum theoretic probabilities require only an expectation assessment of $E(s)$ within the interval $(1.1213, 2]$, rather than support the fabled (but mistaken) value of $2\sqrt{2}$.

Concluding remark

Hardly impossible after all, we have formalised a mathematical structure for the supplementary variable explanation of the results of theoretical quantum physics pertinent to the Aspect/Bell problem. To be sure, we have done this in the specific case of a two-dimensional problem. However, we have recognised in Chapter 2 that claims this cannot be done in a three- or four-dimensional problem are mistaken, merely dragging in a red herring. The higher dimensions involved in expanding the construction to the more complicated design considered by GHSZ (1991) prove no hindrance to formalising more complex partitions of the ensemble of possibilities for the quantity observations they allow. Nor does the 3×3 problem proposed by Mermin (1993) in two dimensions cause a problem, as we shall find in the next chapter.

In any such problem, quantum probabilities are derived for quantity observation possibilities that appear as eigenvalues of specific Hermitian operators acting on some Hilbert space of states. Formally, the implication of the hidden variables explanation of a quantum probability problem is that there can be further as yet unstipulated operators acting on the Hilbert space of an embellished state variable. These would presumably commute with the operators whose stochastics are already prescribed by the theory. Their observation values would provide a functional basis for identifying the quantum observations with greater precision. The specification of such operators could be posed as an abstract mathematical problem. It would be a problem of scientific physics to imagine and design an observable physical process whose engagement could be codified by such operators. Until such progress might be made in experimental research physics, current theory can well be considered to be incomplete.

Chapter 4

MORE HOOJUMS THAN BOOJUMS:
Quantum Mysteries for No one

The detailed mathematical formalities of theoretical quantum mechanics preclude their understanding by even the technically sophisticated among the generally educated public. Replete with measurement operator matrices on a Hilbert space of quantum states, and using a peculiar style of notation that is unique to them, engagement with their prescriptions is forbidding. Aware of the widespread public interest in the inscrutable content of the theory, David Mermin (1971) devised an engagingly simple parable to provide an exhibition of touted features of mysterious quantum behaviour as they have been long understood. Requiring no knowledge of any aspect of quantum physics at all, the exposition merely describes a machine that sends a pair of balls in opposite directions from a central station C to detectors at stations A and B. The balls can address each detector in three different ways, represented by three numbered settings of a dial on its face. Thus, there are nine different conditions under which an experimental run of the machine can be conducted. Coloured lights, either red or green, at the two detectors provide signals as to what occurs in the encounters of the balls at the two stations. Statistical properties of the signal performance of the device in a sequence of operations are presented in such a way as to exhibit one of the defining puzzling mysteries of quantum theory: the purported defiance of Bell's inequality by the probabilistic behaviour of entangled particles.

Questions arise concerning the physical process producing the machine's output. This evidently involves entangled probabilities of light signals at the stations A and B, each of which depends on both the dial setting at its own station and the setting at the other station. This is despite the fact that there is no physical connection between the stations that might convey information between them regarding their respective dial settings. An information transmission scheme is envisioned by which the pair of balls may carry within themselves unobserved encoded messages to stimulate the observed entangled behaviour of the light signals. Although this is shown capable of accounting for regularly matching signals when the dial settings are identical, an enigma arises when the settings are different. Any such scheme appears to instigate matching light signals at the two stations with a frequency exceeding $1/3$ in situations for which the machine is known to exhibit such signalling with probability of only $1/4$. The machine behaviour is touted as mysterious, defying explanation by encoded messages, and portraying one of the great mysteries of quantum analysis.

Upon completion of the exhibition, it is explained to any QM-enlightened readership that the parable of the mysteries actually mimics the situation of a real quantum experiment. This would involve the transmission of a pair of electrons in opposite directions over long distances toward two observation stations at which Stern-Gerlach magnets identify the electromagnetic spin of each electron as directed up or down. The magnets at the two stations can each be set up in any of three differently angled directions perpendicular to the direction of the incoming electrons. These alternative directions are represented in the parable by the three different settings of the dials at stations A and B. The statistics reported in the parable summarising signal behaviour of the machine correspond to what is expected of the spin observations, according to the principles of quantum theory.

The results are both simple and stunning. Mermin advised that he could actually create this machine using the results of paired quantum experiments as the generators of the random outcome sequences. Requiring an effort that he teased would be somewhat less than the order of the Manhattan Project, he proclaimed that "the conundrum posed by the behaviour of the device is no mere analogy, but the atomic world itself, acting at its most perverse."

So engaging, simple, and startling is the exhibition that the piece has become standard fare for the exposition of Bell's inequality to students ever since, even at graduate levels of both physics and philosophy. Moreover, it is included in a welcome and popular collection of essays on matters of theoretical physics meant for the generally educated public, *Boojums all the way through: communicating science in a prosaic age* (Mermin, 1991). Immensely successful and influential, this has been reprinted by now in nine hardcover and five paperback editions. The exposition of "Quantum Mysteries" was lauded by Richard Feynman as "one of the most beautiful papers in physics that I know of", according to the preface to the volume.

The allusion to "boojums" comes from Lewis Carroll's poem, "The Hunting of the Snark", and Mermin describes its emergence as a metaphor for a particular physical phenomenon in his preface to the collection. My own allusion now to "hoojums" comes from memory of my mother's usage, referring to "hoojums and boojums" as quasi-mysterious claims that amount to nonsense. I am not sure, but I think this may have been a common expression among cultured teenagers in the 1930s. At any rate, that is the allusion I make in the title of this chapter. For Mermin's lionisation by another leading figure of twentieth-century physics does not make his analysis correct. I make bold here to show that his argument is both mistaken and misleading. A recognition of my assessment will suggest a revision of physicists' attitudes towards the results of quantum theory, and the mistaken supposed defiance of Bell's inequality in particular.

David Mermin is one of the most accomplished physicists of our era, a cherished professor in the Department of Physics at Cornell University for many years. You would be edified to view his curriculum vitae of publications and activities. In tandem with his professional research, he has been seriously committed to the exposition of the results of theoretical and applied physics to the general public. Along with 138 technical publications, his curriculum vitae includes 20 pedagogical articles and 29 general writings. This does not make him immune to mistakes. We all make mistakes. It is with due respect for his accomplishments, and an appreciation of his expository style, that I explain in this chapter the serious consequences of his mistaken representation of Bell's theorem and its implications. Feeling gauche to refer to him regularly throughout the presentation as "Mermin", I alternately refer to him as "the professor". I intend this with respect.

Of course I invite you to read or to reread his parable of "Quantum Mysteries for Anyone" for yourself, along with the preface to *Boojums*, as a prelude to studying my exposition. Both are now available online. However, to make my presentation self-contained, I will begin with a faithful outline of the mysteries, firstly with a description of the properties of the machine's operation, and then with an exposition of the mysterious behaviour attributed to its conduct. Reflection on the structure of the argument allows us to recognise a sleight of hand in the proposed mimicry of quantum behaviour by the machine's activity. This involves consideration of the gedankenexperiment that underlies the supposed defiance of Bell's inequality. Then we'll address directly the structure of the material problem of quantum physics that the professor would have us ignore, in deference to thinking about his wondrous machinery. The actual problem under consideration is found to involve a system of restrictive functional relations that are neglected in the mechanical parody of quantum behaviour.

We shall then assess a Monte Carlo simulation of the quantum gedankenexperiment that embeds the neglected functional restrictions, displaying a frequency of matching lights on the order of .375 in situations for which Mermin proclaims his machine to provide only .25. The simulation subscribes completely to the probabilities specified by quantum theory in all appropriate instances. Finally, we shall complete a computational analysis of the quantum gedankenexperiment, relying on an application of Bruno de Finetti's fundamental theorem of probability to identify bounds on the relevant probabilities that quantum theory motivates. Quantum theory is quite explicit in refraining from asserting joint probabilities for the outcomes of measurement operators that do not commute. Mathematically speaking, the prescriptions it does motivate do imply bounds on such probabilities that cohere with the explicit assertions it provides.

Mermin's machine

From a box, labelled C in the Figure atop the following page, two apparently indistinguishable balls are ejected in opposite directions toward identical receivers at stations labelled A and B. There are no discernible connections between the components of the machine,

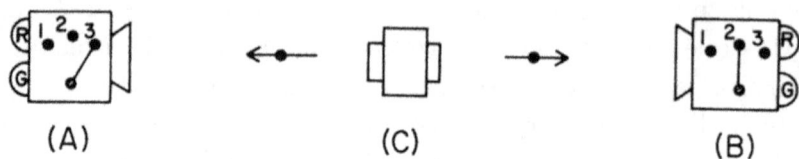

(A) (C) (B)

Fig. 2. The complete device. A and B are the two detectors. C is the box from which the two particles emerge.

Figure 4: The original caricature of Mermin's mysterious machine, reprinted with permission from *The Journal of Philosophy*.

A, B, and C. Each receiver has a dial on its face that can be positioned to any one of three settings, numbered $1, 2$, and 3. Neither receiver is advised of the dial setting on the other receiver. In whatever way this pair of dials are set, when the balls enter the receivers at the stations, each station will flash one of two lights, coloured Green and Red. The results of a sequence of machine experiments are recorded using ordered notation such as $32GR$. This exemplar would designate that an observation was made with the dial at A set to 3 and that at B set to 2, and that the coloured light observed at A is Green while that at B is Red. Thus, a sequence of such observations at various dial settings might look something like $12GG, 31RG, 32RR, 13GR, 22RR, 12RG, 11GG$, and so on.

Here is how the machine works. Pairs of apparently identical balls are sent out in opposite directions from box C toward the two detectors at whatever dial settings have been arranged, and the signal lights are observed. Once the result is recorded at these settings, another pair of balls is sent out to the detectors in whatever next settings are arranged for them. Such experimentation continues sequentially with new pairs of indistinguishable balls, over a long run. When the dials at A and B happen to be set to the same number for any run, then the colours of the flashing lights are always observed to be either both Green or both Red, with equal frequencies of $1/2$ as GG and as RR. On the other hand, when the dials are set at different numbers, the signal lights flash the same colour $1/4$ of the time, and flash different colours $3/4$ of the time. In the former cases, half of the time the identical colours show GG, and half the time they show RR. In the latter cases, half of the time the flashing colours show GR, and half the time they show RG. There is no apparent regularity in the orders of their appearance.

The puzzling question and the source of the mystery involved concerns the determination of what could account for such observable results. It seems odd that the signal behaviour at each detector depends on the dial setting at the other detector, yet there is no obvious way for the two receivers to communicate with one another as to the positions of their dial settings. Proposed as a solution is that while the two balls sent to A and B appear to be identical to one another in every way, the character of each pair may be different in successive runs in a way that is not noticeable to the eye. On any given run, the pair of balls may be somehow encoded with the same one of eight possible labels: $GGG, GGR, GRG, GRR, RGG, RGR, RRG$, or RRR. During a long sequence of runs, the source bin of the pairs of balls provides equal numbers of balls encoded with these eight configurations, in a random order. When either ball from an identically encoded pair such as RGR, for example, enters a detector station, the signal light there would flash Red if the dial at that station were set at 1, would flash Green if the dial were set at 2, and would flash Red if it were at 3. That is, whatever the encoded message on the pair of balls may be, the colour flashed at each detector would match its dial setting with the associated colour designated at that position on the ball's encoding string.

Such a scheme would easily account for the fact that when the identically encoded pair of balls enter the two detectors, the signal lights would always flash the same colour if the detector dials were set to the same number. The balls would be coloured either both Green or both Red, depending on the specific identical encoding of the pair of balls and the setting of the dials as 11, 22, or 33. For example, if the dials were set at 22, a ball encoded with RGR would stimulate the result $22GG$. But what happens if the dials at A and B point to different number settings? Let's consider what is proposed as a mystery.

Mermin's mystery: can you believe it?

Professor Mermin pronounces that if such a scheme were employed, the proportion of runs in which the lights signal the same colour would exceed 1/3 whenever the dials are set differently at the two stations. This would obviously defy the known result that the frequency with which matching lights are observed in such situations equals 1/4. Here is his reasoning. The dial settings at A and B are different from one another in six of the nine paired dial settings (called "case b" settings): $12, 13, 21, 23, 31$, and 32. If the encoded message were RRR or GGG, the signal lights would always shine the same colour when the encoded balls enter the detectors at these settings. Now consider any encoded message that involves two designations of one colour and one of the other, such as RRG or GRG. He writes: "Suppose, for example, that both particles carry the instruction set RRG. Then out of the six possible case b settings, 12 and 21 will result in both detectors flashing the same colour (red), and the remaining four settings, 13, 31, 23, and 32, will flash one red and one green. Thus both detectors will flash the same colour for two of the six possible case b settings. Since the switch settings are completely random, the various case b settings occur with equal frequency. Both detectors will therefore flash the same colour in a third of those runs in which the particles carry the instruction set RRG."

This argument is said to display the mysterious character of the machine. The encoding of the balls would suffice to explain why the flashing lights show the same colour when the dials are set identically at A and B. However, when the dials are set to different numbers, the encoding scheme would seem to imply that at least 1/3 of the observations should exhibit matching colours. Yet the machine is known to produce matching colours in only 1/4 of the runs with such settings. An invisible encoding of the balls seems to contradict the facts of the empirical observations of the machine performance. It appears that the proposal of the hidden encoding cannot account for the facts. No other proposal has been offered that can account for the assured matching light signals whenever the dials happen to be set the same. There must be some mysterious connection between the machine components and the observation process itself to account for the facts.

Would you like to dwell on this puzzle yourself for a while if you have not already done so? Knock yourself out. Literally thousands upon thousands of people have done so, and have been taken in by a sleight of hand in the argument. Without warning or fuss, Professor Mermin has switched the setting of the experimental runs on us. Rather than counting the observed sequence of spin-products as each pair of balls enters the machine at a selected dial setting, he is counting the spin-products for each pair of balls as if it would pass all six of the mixed dial settings. His reported lighting statistics pertains to one ballgame, and his counting of the matching colours pertains to another, two completely different games. As we shall see, it makes sense to consider both games. However, the games are different. As it turns out, they both need to be considered, for different reasons. We require some more thinking.

A magician, not a sleuth: the sleight of hand

It is easy enough to design a simulation for the machine as it was initially described. The pair of balls chosen for the runs of the machine appear identical, without encoding. We can generate colour signals that always appear randomly as GG or RR when the dials are set identically, while they appear in each of these ways only $1/8$ of the time when the dials are set differently. In this latter condition, the colour signals appear as GR and RG each $3/8$ of the time. These results would match Mermin's description of the machine behaviour. The only wonderment involves how the machine system could know the setting numbers at both station A and B so to generate such behaviour in experimental runs, when there is no communication available between its three components. Station C merely sends out pairs of balls, while stations A and B merely set their dials. It is this aspect of the light signal generation that the encoded balls are proposed to resolve.

Simulations involving encoded balls will be considered in two ways: one to mimic the design of the professor's machine, and one to mimic the alternative scheme in which the various encoded ball pairs are sent to all nine dial settings. In the first, we shall randomly pick a sequence of the eight ball encodings, and send the so-encoded pairs to the stations A and B at one of their randomly selected dial settings. The results will not match the operations of the machine at all, as envisaged. In the second simulation, we shall randomly pick an encoding design for each emitted pair of balls, and then engage each pair at all nine dial settings on the receivers sequentially, as he mysteriously assesses them. The light signals at the stations are determined from from the coding rule. Surprisingly, while the second scheme uses the quantum probabilities appropriate to the machine behaviour, it will generate exactly the type of behaviours that the professor denigrates.

Examine firstly the *columns* of Table 1, on the next page. The column components each display the light signal response of the machine to individual pairs of the eight ball encodings at one of the nine dial settings, as described in Mermin's proposal. When the dials are set differently, the light signals are matching for 1/2 of the ball codings, not 1/4. The proportion of matching lights would reduce to 1/3 only if balls coded *GGG* or *RRR* were never introduced to the detectors, while the six mixed encodings were introduced at random. The proportion could reduce to 1/4 only if the distribution of emitted ball encodings varied according to the paired dial settings. They do not. The encoding scheme surely does not generate the behaviour of the machine as it is designed.

Table 1: Encoded messages and their induced responses

The minus symbol (−) designates the display of *matching* light colour signals, and the plus symbol (+) designates *differing* colours, when a row-encoded ball meets a column dial pairing.

Dials Setting	11 1	12 2	13 3	21 4	22 5	23 6	31 7	32 8	33 9
GGG	−	−	−	−	−	−	−	−	−
GGR	−	−	+	−	−	+	+	+	−
GRG	−	+	−	+	−	+	−	+	−
RGG	−	+	+	+	−	−	+	−	−
GRR	−	+	+	+	−	−	+	−	−
RGR	−	+	−	+	−	+	−	+	−
RRG	−	−	+	−	−	+	+	+	−
RRR	−	−	−	−	−	−	−	−	−

However, Mermin, the magician, motivated his claims regarding matching-colour frequencies *among encoded balls* by an argument based on a different situation: experimental results from sending each pair of encoded balls to all nine dial setting pairs. His count of two matching lights among six observations arose from observing each pair of mixed-encoded balls such as RRG as it enters all six distinct station pairings with differing dial settings: 12, 13, 21, 23, 31, and 32. These counts are exemplified in the *rows* of Table 1. To propose that counts from this experiment can represent counts from the original experiment (in which each pair of balls addresses only a single dial setting) amounts to a sleight of hand.

Now who said anything about subjecting a pair of encoded balls to all nine of the paired dial settings? In the operations of the machine, which exhibit 1/4 matching lights when the dial settings differ, each pair of balls is ejected toward a single setting of the dials at stations A and B, and the result might be recorded as something such as $32GR$ or $13RG$. But not both! ... not to speak of results from sending this pair of balls sent to the receivers at the seven other dial pairings. If we would like to study the observed light signals when any single pair of encoded balls is sent to the detectors at all nine paired dial settings, we would require a more elaborate recording structure. There is good reason to make such a study if we are to examine the relevance of the machine behaviour to the touted violation of Bell's inequality by the probabilities of quantum physics, ... and we shall.

The gedankenexperiment with the machine

Suppose we do send each pair of identical encoded balls to all nine paired dial settings, ordered as $11, 12, 13, 21, 22, 23, 31, 32, 33$, and numbered 1 to 9. Designating observations of matching light colours by a -1, and mixed-light-colour observations by $+1$, the experimental results would need to be recorded not merely by something like $13RG$, but rather by something more extensive, such as $(-1, -1, +1, +1, -1, +1, -1, +1, -1)$. Recognise that the vector components 1, 5, and 9 must all equal -1, because the light colours surely match at these settings. Thus, it may seem there would be scope for a sizeable number of distinct observation vectors to arise from such a 9-ply experimental run, perhaps even $64 = 2^6$ of them. For only six of them might each be either -1 or $+1$. As it turns out, the number of possibilities is much smaller. We shall study this detail shortly, when we address the gedankenexperiment of quantum physics that the mysterious machine is proposed to emulate.

In order to assess Mermin's imagined scenario of sending each pair of balls to all nine dial-pairings, we would need to record the machine's light signal response to each encoded pair at every one of the distinct dial settings. The accumulating data matrix would have size $N \times 9$ rather than merely $N \times 1$. Of course, we shall want to use appropriate quantum probabilities when generating such a sequence, and we shall. (These involve matching coloured light probabilities of 1/4 when the two dials are set differently.) When we do this in the context of a physical quantum experiment involving electron spins, surprisingly we shall find that the frequency of quantum observations corresponding to matching lights indeed exceed $1/3$ among the different-dial-setting runs, just as Mermin has identified in his consideration of the encoded balls. The same quantum probabilities that suggest his machine's results of $1/4$ matching light signals in runs on a sequence of balls at a single dial setting, also generate results of matching signal frequencies exceeding $1/3$ when each pair of balls engages all nine settings. There is nothing mysterious about it. What has been missed in Mermin's accounting are the same type of functional relations among the spin-products in a gedankenexperiment that Aspect/Bell missed in their simpler polarisation experiment with paired photons.

Now, there is another peculiarity to be noticed in the results of Table 1. Follow the professor across a row for any mixed-colour-encoded ball pair, and notice with him the two of six matching light colour results when the two dials are set differently. However, notice additionally that there are only *four distinct rows* of nine-vectors that can possibly result from the scheme using the eight types of ball encodings. There are eight rows to the Table, but the final four rows duplicate the first four rows! Rows 5, 6, 7, and 8 duplicate rows 4, 3, 2, and 1. Just two paragraphs ago we were imagining the possibility of several possible nine-vector observational results in the second scenario of machine operation, as many as 64. Now it is evident that there are only four! The vector $(-1, -1, +1, +1, -1, +1, -1, +1, -1)$, which we suggested as an exemplary possibility, would not be a possibility at all under the encoded-ball scheme, despite it exhibiting matching lights in two of the six paired dial settings that differ. Only four possible nine-vectors of result possibilities arise from the eight encoding designs. The bottom line for now is that this scheme of sending encoded balls to detectors at all nine dial settings is a proposition completely different from that which yields the proclaimed results of Mermin's machine. We shall sort this all out forthwith.

To catch the magician at his game requires a diversion into the real quantum experiment that the professor would have us ignore, while we are enticed to marvel at his mysterious machine. My plan is to begin with a presentation of the relevant practical quantum experiment that can be and has been conducted many times. Then we shall embellish the context to a gedankenexperiment designed to assess both the implications of Einstein's principle of local realism and his challenge to the completeness of quantum theory. This is the context in which the specification of Bell's inequality is entertained, and the context for which Mermin's second version of the ballgame is appropriate as an emulation. Only when we recognise the structure of this matter will we be able to identify prospective quantum probabilities appropriate to a single pair of balls visiting all nine designs of dial settings.

Having studied the structure of the real quantum experiment and its associated gedankenexperiment, we shall design and conduct a Monte Carlo experiment as a prelude to a complete analysis of the entire situation, based strictly on the limited claims of quantum theory. As with our solution to the simpler experimental context of Aspect/Bell in Chapter 1, we shall find that quantum theory quite explicitly does *not* propose a joint probability distribution over the complete space of possible gedanken observations. Rather, in its current incomplete form, it specifies a multi-dimensional polytope of such distributions. Furthermore, it renounces any prospect for refining it.

The gedankenexperiment sending the same pair of balls to all nine dial pairings allows a Monte Carlo simulation that emulates a simple design conforming to the probabilities of quantum physics. Surprisingly, it also exhibits matching light frequencies exceeding 1/3 under conditions that Mermin proposes as mysterious. But the Monte Carlo simulation will not constitute the end of our analysis. The simulation design, while natural, is not the only design that QM theory would allow, and we shall see why. This is all sounding complex. However, it is merely a matter of plodding on. Let's go!

What are we really talking about?

Professor Mermin understands the mystery to convey that while there are no obvious physical connections between the three pieces of the experimental device, A, B, and C, the attempt to explain its experimental features by unobservable instructions encoded within the balls is futile. Such an explanation might account for the specified observable outcomes of the machine runs with dials at A and B set identically as $11, 22$, or 33, but it seems to provoke specific observations of flashing lights that do not match what we experience when running the machine at other dial pairings. The alternative he proposes is to recognise that indeed the operation of the recorders actually are connected in some mysterious way. However, these would be "connections of no known description, that serve no purpose other than relieving us of the task of accounting for the behaviour of the device in their absence." This is the purportedly mysterious behaviour of quantum mechanics. Yet he engages such speculations no further, as the task proposed for his exposition was merely to state the conundrum, not to resolve it. The parable is concluded.

After completing his description of the mystery, the professor presents an insightful discussion of the relevance of the parable to issues raised by Einstein, Poldolsky, and Rosen (EPR, 1935) in their proposition that the theory of quantum mechanics must be incomplete. While they had presented arguments that may appear telling regarding the activity of the machinery when the dials are set *identically* at A and B, their arguments appear to fail in situations in which the dials are set differently. This was a situation they did not assess, consumed as they were with claims about the reality of quantum states and experimental observations that could be predicted with certainty. The implications for quantum behaviour portrayed in the parable by different dial settings at A and B did not become evident until the startling research results of John Bell. These have been understood to certify that if one presumes Einstein's principle of local realism, then the specifications of quantum theory defy some standard inequalities of probability theory. Although we have debunked this in Chapter 1, this has been long the accepted state of affairs. If you read the history of the quantum theorists' interactions throughout the twentieth century in the fascinating account of Adam Becker (2018), you will find the same claims as those of Mermin's mysteries. They appear only in a slightly different guise in his account of Bell's contribution to developments. We shall clarify the entire matter in what follows.

Mermin's exposition concludes with a description of the contextual quantum experiment that the parable is meant to portray, emphasising that such detail can be conveniently ignored while the significance of the mystery is absorbed in awe. This is a well-known quantum experiment involving a pair of electrons that are propelled in opposite directions toward identical detecting devices of Stern-Gerlach magnets at stations A and B. Each magnet is oriented at one of three specific angles within the plane perpendicular to the incoming electrons. The detectors identify the spin directions of the pair, each in either the direction "up" or "down", denoted by

$A = +1$ or $A = -1$, and similarly for the value of B. Rather than ignoring the physics experiment as suggested, the remainder of my exposition now is directed to a detailed assessment of the exact specification of this experiment and the proclamations of quantum theory that concern it. We shall find that Mermin's parable fails to represent the situation adequately, for the same reason that Aspect's assessment of Bell's inequality fails in the simpler case of a pair of photons presented to two paired polarisation angles. It is a mathematical error of neglect.

The quantum experiment in gedanken extension

Remember that the simple quantum gedankenexperiment of Aspect/Bell concerned a cerebral assessment of possibilities for the combined result of four practical experiments, each of which can be engaged, but for which the engagement of all four simultaneously is recognised as impossible. The same style of investigation will pertain to our considerations now. We shall examine the (im)possible imagined results of a nine-ply Stern-Gerlach thought experiment, conducted on a single pair of electrons. This is the context to which Bell's inequality pertains, and in which the principle of local realism is relevant. The pair are sent in opposite directions toward the activity stations A and B, monitored by Alice and Bob. In a real experiment, each of them is charged with observing one component of the pair as it passes a magnet oriented in one of three different directions relative to vertical up and down. The vertical position is designated as the zero position, and the other two are directed in twists of negative and positive angles relative to this zero. The possible pairings of these magnet orientations at the two stations specify 3×3 possibilities for a paired choice of them at the two stations during any experimental run. The vector outcome of *nine* spin-products occurring in a run of a thought experiment sending a single electron pair to all nine paired magnet orientations will be denoted by $\mathbf{G}_9$, the "G" standing for "gedankenvector".

Initially, we shall designate the three possible magnet orientations of each spin observation variable by the subscripts n, z, or p, so as to represent its alignment relative to vertical at an angle that is negative, zero, or positive. We may write A_p or B_z, for example. When referring to a spin observation at a generic magnet direction, we may we may use a subscript letter "d" for the direction, considered as a variable equal to either n, z, or p. Sometimes I shall refer casually to the quantities A and B without any subscript. Eventually, we shall assess a specific setup design in which the two chosen magnet orientations *differ* by the angles $\theta = -120°, \theta = 0°$, and $\theta = +120°$. This is the setup relevant to the probability assessments prescribed in Mermin's parable. Deliberations of quantum theory specify probabilities for the possible paired observations of spin values at the two stations, denoted as P_{++}, P_{+-}, P_{-+}, and P_{--} appropriate to a specific paired angle setup. Equivalently, it specifies the expectation of the spin-product, $E(AB)$.

Although the structural similarities of the experiments on paired photons and on paired electrons are apparent, the quantum probabilities determined for the possible paired spins that might be observed are different in their details. For the record, the quantum probabilities pertinent to paired electron spins for which the relative angle between their Stern-Gerlach magnets is θ are as follows.

Joint spin probabilities:

$$P[(A_d = +1)(B_d = +1)|\theta] = P[(A_d = -1)(B_d = -1)|\theta] = \tfrac{1}{2} sin^2(\theta/2) ,$$

and

$$P[(A_d = +1)(B_d = -1)|\theta] = P[(A_d = -1)(B_d = +1)|\theta] = \tfrac{1}{2} cos^2(\theta/2) .$$

Marginal spin probabilities:

$$P(A_d = +1) = P(B_d = +1) = 1/2.$$

and Conditional spin probabilities:

$$P[(A_d = +1)|(B_d = +1), \theta] = sin^2(\theta/2) \neq P(A_d = +1) = 1/2 ,$$

and

$$P[(A_d = +1)|(B_d = -1), \theta] = cos^2(\theta/2) ,$$

which is different still. Further relevant is the

Expectation of spin-product observations: $E[A_d B_d|\theta] = -cos(\theta).$

In the gedankenexperiment, Alice and Bob will observe the spins of a pair of electrons in *every* paired directional setting of their magnets. Their respective observations named A and B would be recorded as either $+1$ or -1 to designate an observation of spin "up" or "down". We shall denote the individual components of possible results of their nine paired observations as product events, such as $(A_n = +1)(B_z = -1)$ or $(A_z = -1)(B_p = -1)$ and so on. My use of arithmetic notation means that each of these *product events* indicates whether these *joint* observation of spin values at site A and site B arises in a particular configuration or not. The spin-products themselves would be $A_n B_z = -1$ and $A_z B_p = +1$ for these examples, respectively. There will be nine of them for any run of the gedankenexperiment. The number of *prospective* nine-tuples of observation pairs could be as large as $2^9 = 512$. Shortly, this number of possibilities will be found to be much smaller, both on account of theoretical speculation and the particular directional angles we employ in the experiment's design.

We shall begin by considering a list of all the possible results of the paired observations at A and B that could be entertained according to Einstein's contentious (and currently widely rejected) "locality" condition. This involves a proposition that lies outside the technical domain of quantum theory. It accepts that the electron spin observation made by Alice in any specific magnet orientation is a result assessed with a quantum probability entangled with that of Bob's magnet direction in that instance. However, it proclaims that when Alice's magnet orientation is set in any specific experimental run, her spin observation in this instance would be the same (either up or down) at every corresponding orientation of Bob's magnet, and no matter what might be his spin observations in these imagined companion experiments *on the same pair* of electrons. For the behaviour Alice actually observes in any specific instance presumably depends only on the local conditions at her station during the run.

The reason such a claim lies outside the scope of quantum theory is that the distinct operator matrices specifying observations of a single pair of electrons addressing two different paired magnet orientations do not commute. Thus, quantum theory itself says nothing about their *joint* spin-product results at the two designs. The complete thought experiment presumes that a single pair of electrons passes by the two magnets in all nine of their paired orientations. The principle of local realism stipulates that if Alice's spin observation is, say, $+1$ in a specific magnet orientation when Bob's relative orientation is $+120°$, then Alice's would also equal $+1$ in this instance if Bob's magnet were oriented relatively at $0°$ and/or at $-120°$ as well. Bob's spin observations might be either $+1$ or -1 in either case. Although Alice's actual measurement for a particular electron spin is proposed to be invariant with respect to the setting of Bob's detection angle, this principle respects nonetheless an assertion of entanglement of the electrons. The entanglement is identified via the specification $P[(A = +1)(B = -1)|\theta] = \frac{1}{2}\cos^2(\theta/2)$

at any single relative angle setting. Here θ is the relative angle between Alice's and Bob's magnet orientations. Equivalently, the specification is $E(AB|\theta) = 1 - 2cos^2(\theta/2) = -cos(\theta)$. These are the relevant prescriptions of quantum theory pertinent to paired electron spins.

The principle of local realism implies that in measuring the spins at all nine angle orientation pairs for the gedankenexperiment, *each* of the investigators would observe *only three distinct* spin values. According to this premise, the observed values of A_n, A_z, and A_p in the nine-ply experiment would each arise as the same value, no matter which of the station B magnet orientations it were paired with, B_n, B_z, or B_p. The same would hold for the observation values of the B's paired with A's. These six observation values would display themselves among nine specific observation pairs, each of the form (A_{d_A}, B_{d_B}). There appear to be only $2^6 = 64$ conceivable instantiations of these six spin observations that would respect the principle of local realism in any run of the gedankenexperiment, rather than the 2^9 that we had initially entertained.

However, the probabilistic assertions of quantum theory reduce the number of these possibilities still further. Consider the prescription of quantum theory pertinent to a any experimental setup in which Alice's and Bob's magnet orientations are identical, and we measure the spin values (A_n, B_n), or (A_z, B_z), or (A_p, B_p). Quantum theory stipulates that we must observe opposite spin values at stations A and B in any such experiment. For it specifies that $P(A_d B_d = -1|\theta = 0) = 1$, or equivalently $E[A_d B_d|\theta = 0] = -1$ whenever the two magnet orientations are identical. It is impossible for the spin observations at Alice's and Bob's stations to be the same when their magnet orientations are the same. An implication of this quantum prescription, joined with the principle of local realism, is that there are not 64 possible results of the gedankenexperiment, but rather only 8. We examine them now.

Possible results of the 3×3 gedankenexperiment

Let's cut to the chase, and display on the following page a realm matrix showing all possible results of the 3×3 gedankenexperiment that might be observed for a single pair of electrons passing the two magnets in all of their possible relative orientations. The possibilities constituting the experimental ensemble are presumed to be limited only by the principle of local realism and the assertions of quantum theory, specifically as they are relevant to component experiments in which the orientations of Alice's and Bob's magnets are identical. The matrix is partitioned vertically into three sections and horizontally into two, for reasons we now explain.

Within the first section of the named quantity vector at left, there appear six possible observations of Alice and Bob in the performance of the nine experiments: $\mathbf{G}_6 = (A_n, A_z, A_p, B_n, B_z, B_p)^T$. While each component of the 6×1 vector can equal either -1 or $+1$, the components of the second triple in any column, B_n, B_z, and B_p, must be the negative values of the first three components of that column. Each matching spin-product must equal -1. This is a prescription of quantum theory. For when the directions of the two magnets are identical, then the relative angle beween them is $\theta = 0$. As for their expectation, $E(A_d B_d | \theta = 0) = -cos(\theta) = -1$.

$$
\mathbf{R}\begin{pmatrix} A_n \\ A_z \\ A_p \\ B_n \\ B_z \\ B_p \\ {*}{*}{*}{*} \\ A_n B_n \\ A_n B_z \\ A_n B_p \\ A_z B_n \\ A_z B_z \\ A_z B_p \\ A_p B_n \\ A_p B_z \\ A_p B_p \\ {*}{*}{*}{*} \\ {}^1 A_n B_n \\ {}^2 A_n B_z \\ {}^3 A_n B_p \\ {}^6 A_z B_p \end{pmatrix} = \left(\begin{array}{rrrrcrrrr}
1 & 1 & 1 & 1 & * & -1 & -1 & -1 & -1 \\
1 & 1 & -1 & -1 & * & 1 & 1 & -1 & -1 \\
1 & -1 & 1 & -1 & * & 1 & -1 & 1 & -1 \\
-1 & -1 & -1 & -1 & * & 1 & 1 & 1 & 1 \\
-1 & -1 & 1 & 1 & * & -1 & -1 & 1 & 1 \\
-1 & 1 & -1 & 1 & * & -1 & 1 & -1 & 1 \\
* & * & * & * & * & * & * & * & * \\
-1 & -1 & -1 & -1 & * & -1 & -1 & -1 & -1 \\
-1 & -1 & 1 & 1 & * & 1 & 1 & -1 & -1 \\
-1 & 1 & -1 & 1 & * & 1 & -1 & 1 & -1 \\
-1 & -1 & 1 & 1 & * & 1 & 1 & -1 & -1 \\
-1 & -1 & -1 & -1 & * & -1 & -1 & -1 & -1 \\
-1 & 1 & 1 & -1 & * & -1 & 1 & 1 & -1 \\
-1 & 1 & -1 & 1 & * & 1 & -1 & 1 & -1 \\
-1 & 1 & 1 & -1 & * & -1 & 1 & 1 & -1 \\
-1 & -1 & -1 & -1 & * & -1 & -1 & -1 & -1 \\
* & * & * & * & * & * & * & * & * \\
-1 & -1 & -1 & -1 & * & -1 & -1 & -1 & -1 \\
-1 & -1 & 1 & 1 & * & 1 & 1 & -1 & -1 \\
-1 & 1 & -1 & 1 & * & 1 & -1 & 1 & -1 \\
-1 & 1 & 1 & -1 & * & -1 & 1 & 1 & -1
\end{array}\right).
$$

By contrast, when the magnet orientations of Alice and Bob differ, then a spin observation at A in either direction might be accompanied by a spin observation at B in either direction as well. The spin-product might equal -1 or $+1$. This feature is evident in rows 2, 3, 4 and 6, 7, 8 of the middle partitioned section. These exhibit the arithmetical *products* of each of the three A's with each of the three B's appearing in the column. The nine components of the second partition of the quantity name vector are

$$(A_n B_n, A_n B_z, A_n B_p, A_z B_n, A_z B_z, A_z B_p, A_p B_n, A_p B_z, A_p B_p)^T.$$

The third partition of the named quantity vector and of the realm matrix merely repeat rows $1, 2, 3$, and 6 of the middle partition. I shall discuss them when it becomes appropriate. We are ready now to examine the *vertical* partition of the displayed realm matrix.

Functional relations among spin-products

A *substantive* matter to recognise about this realm matrix is that the four columns of the right partition of the middle section constitute a *reversed replica of the columns of the left partition* of that section. The middle section of column 5 is identical to that of column 4. That of 6 is identical to column 3, and so on, until the midsection of column 8 is identical to that of column 1. Thus, the middle horizontal partition of the realm matrix has only 4 *distinct* columns, rather than 8. In contrast, columns of the top partition above it are distinct. What is more, notice that the middle partition matrix has only four distinct rows as well! This fact has important ramifications for the analysis of the hypothetical problem of nine distinct Stern-Gerlach experiments on the same pair of electrons.

In the first place, the realm matrix for the spin-product vector possibilities resulting from the gedankenexperiment on a single pair of electrons consists only of the left half of the middle partition matrix above, a 9×4 matrix. For reference in ensuing discussions, we shall refer to this realm matrix for the vector of nine spin-product results of the gedankenexperiment as $\mathbf{R}_{9,4}$. What is more, some rows of this complete matrix are obtainable via specific functions of other rows in that section. As long as any two rows of $\mathbf{R}_{9,4}$ constitute the Cartesian product $\{-1, +1\}^2 = \{(-1, -1), (-1, +1), (+1, -1), (+1, +1)\}$, the remaining seven rows are determined by a function of them. This is an "into" mapping with the structure $\{-1, +1\}^2 \to \{-1, +1\}^7$. Consider, for example, the rows 2 and 3 of $\mathbf{R}_{9,4}$. Their column pairs exhaust the Cartesian product $\{-1, +1\}^2$, constituting the domain of a function. For each of these pairs in the domain, the remaining seven columns provide a unique vector within $\{-1, +1\}^7$. We shall designate this function with the notation $23 \to 1456789$.

In fact there are twelve such functional relations that inhere within the structure of the realm matrix $\mathbf{R}_{9,4}$. Using the same functional notation, we can list them as:

$23 \rightarrow 1456789$ $34 \rightarrow 1256789$ $47 \rightarrow 1235689$

$26 \rightarrow 1345789$ $36 \rightarrow 1245789$ $48 \rightarrow 1235679$

$27 \rightarrow 1345689$ $38 \rightarrow 1245679$ $67 \rightarrow 1234589$

$28 \rightarrow 1345679$ $46 \rightarrow 1235789$ $78 \rightarrow 1234569$.

The first of these arrows denotes the functional relation among the columns of spin-products we have just described. It is a mapping from $\{-1, +1\}^2$ into $\{-1, +1\}^7$. The subsequent arrow structures constitute an exhaustive list of all inhering functional relations. The sole functionality requirement is that the rows of $\mathbf{R}_{9,4}$ corresponding to the domain variables are the component vectors of $\{-1, +1\}^2$.

There are six paired spin-products corresponding to experiments with different magnet orientations at the stations of Alice and Bob. These are the components 2, 3, 4, 6, 7, and 8 of the spin-product vectors. There are $^6C_2 = 15$ possible choices of two spin-products to consider as elements of a possible function domain. However, three such choices of two of them fail to constitute a function domain: (2,4), (3,7), and (6,8). That is, attempted mappings of 24 onto 1356789, of 37 onto 245689, and of 68 onto 1234579 all fail to identify a function. For example, the column pairs from rows 2 and 4 do not exhaust $\{-1, +1\}^2$. Furthermore, when they repeat they would specify two distinct mapping objects in $\{-1, +1\}^7$. The mapping they provide for examination does not constitute a function.

A second matter worthy of note is that these embedded functional relations are not linear. If they were, the rank of the middle partition of the realm matrix would be only two, but it is four! This is a feature crucial to the implications of quantum theory for assessing the results of the gedankenexperiment. Quantum theory makes specific expectation (and probability) assertions regarding the spin-products for any pair of domain variables among our listed functions, considered as distinct isolated experiments. If the functional relations were linear, then these would imply precise expectations for the range variables. As it is, the assertions of quantum theory will stipulate only bounds on the expectations for the spin-products of the range variables in any case.

This situation provokes some intrigue. If we were to consider the specification of a joint distribution for the nine component spin-product results of the gedankexperiment, the conditional distribution for any seven products in a range vector, given the results of their domain vector, would be degenerate at their function value. Yet their joint probability distribution with the domain variables cannot be determined precisely. We shall see more of this.

A further remark of note concerns the repetitions found among the column pairs comprising rows 2 and 4, rows 3 and 7, and rows 6 and 8 in this realm matrix. These equalities exhibit an implication of local realism for the spin-product vector of a gedankenexperiment run: the commutativity of spin-product observations with respect to the orientations of their detecting magnets. Noting the names of the spin-product quantities whose observation possibilities constitute the repeating rows, these repetitions specify equality of the spin-products $A_n B_z = A_z B_n$, $A_n B_p = A_p B_n$, and $A_z B_p = A_p B_z$ in any imagined run of the experiment. This feature of local realism will come to bear on our computations of probability bounds for gedankenresults deriving from the claims of quantum theory, restricted by its avowed uncertainty principle.

A final surprise can be seen among the complete columns of the partition matrix $\mathbf{R}_{9,4}$. Although we have mentioned nothing at all about encoded balls while constructing it, it is apparent that the columns of $\mathbf{R}_{9,4}$ match the row designations in Table 1 (here page 106), which repeat themselves. These, remember, exhibit the light signal responses of Mermin's mysterious machine to his suggested explanation based on encoded balls. We shall revisit this recognition too in due time.

We are ready to be entertained by the results of a simulation experiment, to which we shall now turn. I can only advise you that there is still more intriguing detail that can be exposed regarding a nine-fold gedankenexperiment. It pertains to the difference between the structure of the EPR experiment in which paired quantum behaviours were completely informative about one another, and Bell's analysis which concerns paired experiments for which an observation at one station does not provide definitive information about the result observed at the other. At the risk of losing every one of my patient readers, I have deferred a commentary to an appendix to this chapter. Let's now have some fun!

A simulated gedankenexperiment using QM motivated probabilities

We can capitalise on our recognition of the functional relations embedded in the realm matrix $\mathbf{R}_{9,4}$, by conducting a simulation experiment appropriate to the gedankenexperiment on paired electron spins to which Bell's inequality pertains. It represents the situation Professor Mermin was assessing when he evaluated the behaviour of his mysterious machine in response to colour-encoded balls at all nine dial-pair combinations. Our simulation is meant to elucidate the structure that the theory of quantum mechanics designates as appropriate to the assessment of Bell's inequality in the context of this gedankenexperiment. It will generate a sequence of twelve million gedankenvectors $\mathbf{G}_9$, these being simulated spin-products of electron pairs that each pass all nine paired magnet orientations, obeying the probabilistic prescriptions of quantum theory.

The experiment proceeds in twelve formats, corresponding to the twelve functional relations among the gedanken spin-products we have just identified. Data are generated that correspond to a gedankenexperiment of Stern-Gerlach apparatus with magnet orientation possibilities for both Alice and Bob, each set at three orientations relative to vertical in the (x, y) plane: $-120°$, $0°$, and $+120°$. This plane of directions is perpendicular to the direction of the incident electrons, exactly as described in the conclusion to the professor's "Quantum Mysteries for Anyone". We shall denote the relative angle between Alice's and Bob's magnet orientations in any paired direction setup by $\theta = dir_B - dir_A$. The nine paired direction possibilities yield orientations that differ by angles of $\theta = 0°$ when the A and B observations are (A_n, B_n), (A_z, B_z), and (A_p, B_p); of $\theta = +120° = -240°$ when the A and B observations are $(A_n, B_z), (A_z, B_p)$, and (A_p, B_n); and of $\theta = -120° = +240°$ when the A and B observations are $(A_n, B_p), (A_z, B_n)$, and (A_p, B_z). Rows 1, 5, and 9 of our realm matrix $\mathbf{R}_{9,4}$ correspond to an orientation difference angle $\theta = 0°$; row numbers 4, 8, and 3 pertain to the difference angle $\theta = -120°$; and rows 7, 2, and 6 pertain to the difference angle $\theta = +120°$.

The simulation begins with a routine pertinent to the spin-product function $23 \rightarrow 1456789$. It first generates observation values for components 2 and 3 of the spin-product vector independently according to standard QM specifications of probabilities for differing spin observations at these magnet orientation pairings. These prescribe the quantum probabilities

$$P[A_n B_z = -1 \mid \theta = -120°] = 1/4 = P[A_n B_p = -1 \mid \theta = +120°],$$

which correspond to the frequencies reported by Mermin in the observations of his machine. The coloured lights match only $1/4$ of the time when the stations' magnet orientations differ in these ways.

Appropriately then, from each eventuality of the two outcome values so generated for domain components 2 and 3, the associated range values for components 1, 4, 5, 6, 7, 8, 9 are computed according to the functional rule we have designated by $23 \rightarrow 1456789$. In each of these cases, the conditional distribution of these latter component values, given the spin-product pair simulated for the components of the domain, is degenerate on their function value. This derives from the structural features of the possible spin-product observations. From one million such generations, the number of occurrences of -1 are counted and recorded in the spin-product columns 1 to 9. The results of the simulation yield

23 1000000 250191 250332 250191 1000000 625225 250332 625225 1000000 1456789.

The number at the far left edge of this row of such counts designates the row numbers of function *domain* observations that were selected by the pseudo-random numbers of MATLAB, while the number at the far right edge identifies the corresponding rows in the *range* that were computed via appropriate function rules. Understanding this, you should read this report line as identifying 1000000 counts of spin-product observations -1 in experiment columns 1, 5, and 9. (Mermin's lights always flash the same colour when the dials at A and B read the same.) Counts of -1 amounted to 250191, 250332, 250191, and 250332 as recorded in experiments 2, 3, 4, and 7, respectively, two pairs of which involve repetitions. Finally, identical counts of 625225 were recorded in both columns 6 and 8. (The identical counts in three of the paired columns result from the principle of local realism. We found this to insist that $A_n B_z = A_z B_n$, that $A z B_p = A_p B_z$, and that $A_p B_n = A_n B_p$ when the experiments run gedankenly in tandem. This feature arises naturally in the function-based generations of the simulation.)

In particular, notice that the repeated count of 625225 corresponding to spin-products $A_z B_p$ and $A_p B_z$ differs markedly from the claims of Professor Mermin that the experiment should yield a number near to 250000 in *every* spin-product column for which the difference in orientation angle is $\theta = -120°$ or $\theta = +120°$. We can now understand why this has occurred! On account of the functional relations required among the spin-product observations, only a select two of the spin-product values can be generated freely according to QM probability relations as specified by the well-known cosine squared equations,

$$P[(A = -1)(B = +1)|\theta] = \tfrac{1}{2}\cos^2(\theta/2).$$

The remaining seven must be determined from the values of these two according to the functional relation we are considering, which binds them all.

Now there is nothing special about the experimental components 2 and 3, which we allowed to be chosen freely by their quantum probabilities. We have seen that there are twelve such domain choices of experimental pairs that can be used to generate the gedanken spin vector. We shall now examine a *full array* of simulation results deriving the other eleven choices of the function domain as well. Each domain pair will be chosen to begin one million simulation runs of the machine, just as the domain pair 23 was used to start the one above. The results will be found to allay misdirected concerns aired in the parable regarding proportions of matching light colours proclaimed to differ from 1/4. Such concerns *are* appropriate in real physical experiments in which sequences of electron pairs engage any single pair of differing magnet orientations. But they *are not* appropriate to a gedankenexperiment motivated by quantum theory and local realism, in which each pair passes all nine magnet orientation pairings. Let's now examine the complete results that initiate the simulation with the paired spin-products for each of the twelve function domains. These appear in Table 2 on the following page.

Each row of these Simulation Counts is based upon quantum probabilities relevant to the domain of a distinct function that is displayed in its left edge column. Columns numbered 1, 5, and 9 (not counting the left-edge column) show that the generated spin-product is negative at all three configurations in which magnet orientations at the sites A and B are identical. The products of the simulated spin observations yield counts of 1000000 for the value -1 in these columns, exactly as specified by the quantum probabilities. (Mermin's coloured lights are identical.) The negative spin-product counts at the other magnet pairing configurations vary.

Table 2:

Twelve Simulation Counts of Negative spin-products

D	A_nB_n	A_nB_z	A_nB_p	A_zB_n	A_zB_z	A_zB_p	A_pB_n	A_pB_z	A_pB_p	Rng
23	1000000	250191	250332	250191	1000000	625225	250332	625225	1000000	1456789
26	1000000	249641	625501	249641	1000000	249912	625501	249912	1000000	1345789
27	1000000	250096	250274	250096	1000000	624192	250274	624192	1000000	1345689
28	1000000	250188	625260	250188	1000000	250060	625260	250060	1000000	1345679
34	1000000	250777	250397	250777	1000000	624338	250397	624338	1000000	1256789
36	1000000	625459	249849	625459	1000000	249814	249849	249814	1000000	1245789
38	1000000	624890	250619	624890	1000000	250277	250619	250277	1000000	1245679
46	1000000	250093	624872	250093	1000000	249855	624872	249855	1000000	1235789
47	1000000	249640	249716	249640	1000000	625256	249716	625256	1000000	1235689
48	1000000	249483	625658	249483	1000000	249411	625658	249411	1000000	1235679
67	1000000	625506	249710	625506	1000000	249974	249710	249974	1000000	1234589
78	1000000	625491	249736	625491	1000000	249681	249736	249681	1000000	1234569

Sum Simulation Counts by Product Column:

12000000 4501455 4501924 4501455 12000000 4497995 4501924 4497995 12000000

Proportions Differing Spins by Product Column:

1.0000 .3751 .3752 .3751 1.0000 .3748 .3752 .3748 1.0000

The rows of this output display results of twelve (million) simulation runs that are structurally identical in their generation to
those I have already described for the function 23 → 1456789.
Distinct runs were based on quantum probability assertions applied to the two domain variables of each of the twelve functional
relations we have recognised. In each set of runs, spin-products
were generated independently for two appropriate function domain
variables using QM probability simulations. Then the remaining
seven were computed from these using the relevant function specifications we have identified. Following this procedure, the output of
each 9-vector generated respects the injunctions of all twelve function rules. The same reporting format is used as in the introductory
presentation of the first row, described on the previous page.

For example, row 3 of the output, which identifies in its two edge
columns the relevant functional relation as 27 → 1345689, exhibits
counts of 250096 and 250274 in matrix columns (2 and 4) and (3

and 7) respectively, on the order expected according the probability specifications of quantum theory applied to a single experimental magnet settings. However, in columns 6 and 8 of row 3, among the range variables of the constraining function both counts are found to equal 624192, not near to 25000 at all!

Similar structures pervade the simulation counts in all rows of the count matrix: two of the column elements of each row exhibit identical counts in the vicinity of 625000 while four column elements are in the vicinity of 250000, displaying two identical count pairs. No matter which pair of function domains is used to generate the nine columns of results, the counts are identical in columns 2 and 4, in columns 6 and 8, and in columns 3 and 7. The requirement of the principle of local realism is satisfied by recognition of the functional relations among spin-products.

Summing the twelve columns of these simulation counts (which each arise from 1 million simulated Stern-Gerlach experiments) yields further results of interest. Particularly striking are the implied *proportions* of differing spin-products exhibited at the several paired angle orientations. These proportions are *not* displayed as three 1's and six .25's as proclaimed by Professor Mermin. The three 1's surely appear in the expected places, but of the remaining six columns we find *all* the proportions near to .375, defying his claim to the proportion arising as $1/4$. Indeed, the proportions we have generated in the quantum gedanken simulation are reminiscent of the frequency behaviour of encoded balls exceeding $1/3$. This had worried him in his parable, motivating him (with many many others) to decry the sensibility of Einstein's suggestion of hidden variables in quantum behaviour.

Comments, a qualification, and a query

Don't get me wrong. If you would do a sequence of simulation experiments at a specific pairing of differing magnet orientations using quantum probabilities, you would find the spin-product values to equal -1 in close to $1/4$ of these cases. However, if you do a long sequence of simulated experiments that gedankenly subjects the electrons to all nine paired magnet angle directions in the way local realism restricts them, you would find the proportion of spin-products equal to -1 at about .375 whenever the relative angle between the magnets equals $-120°$ or $+120°$. This happenstance governs the counts displayed in columns $2, 3, 4, 6, 7,$ and 8. The result has nothing to do with Mermin's proposed explanation of "the mystery" involving colour-encoded balls. It derives from a recognition of the functional relations embedded into spin-product possibility vectors in the gedankenexperiment. A situation clearly distinct, when the rows are produced by gedankenly submitting each pair of electrons to all nine of the relative angle settings, many elements of the Cartesian product $\{-1,+1\}^6$ for the unequal magnet angle designs would constitute impossible outcomes of the spin-product functions that govern the experiment. Each of the allowable result vectors respects twelve functional mappings of $\{-1,+1\}^2$ into $\{-1,+1\}^7$.

Professor Mermin's fabulous machinery produces no mysterious results at all. The character of the matrix of results would be different, depending on which of the two different ways that the results are generated. This is not surprising.

One way to simulate the gedankenexperiment as supported by the probabilities of quantum theory would be to pick sequentially (randomly, uniformly) one of the functional relations that bind the spin-products $\{-1, +1\}^2 \rightarrow \{-1, +1\}^7$, to generate a vector of nine-tuple observations as we have described. Then pick a functional relation again to generate the next 9-vector of results. Continuing with this process we would generate a sequence of such 9-vectors, and accumulate the counts of negative spin-products at the nine angle pairings across the sequential generations. This process would result in proportions of negative spin-products as appear in the final line of Table 3. However, this result could hardly be claimed to be a definitive prognostication of quantum theory relevant to the gedankenexperiment. Be aware that the nine component vector results of the experimental runs involve simultaneous results of observations at angle pairings whose operator matrices do not commute! Quantum theory explicitly says nothing specific about such impossible experimental results.

So what *does* quantum theory say about the gedankenexperiment, and how can we present it in a complete and concise way? Consider again a single functional relation such as $23 \rightarrow 1456789$. Well, quantum theory is quite specific in identifying a probabilistic structure governing the spin-product result at either of the two relative angles between the Stern-Gerlach magnets that appear in the function domain. However, it is also quite explicit in denying any motivation for making claims about *simultaneous* spin-product results at these two relative angle settings for which the observation matrix operators do not commute. The results of this simulation activity I have just proposed *could be* admissible according to the logic of quantum theory, but there is no reasoning that would make the joint distribution they imply definitive. There is no requirement that spin-products in any function domain be picked independently, and no requirement that the function domains be picked randomly and uniformly at all, as I have done here. We could generate other distributions of results if we picked them according to some other random scheme.

The way to *characterise* the *complete* space of joint probability distributions over gedanken results that cohere with the positive claims of quantum theory is to assess a battery of linear programming computations. These identify the *bounds* on probabilities for the range settings of the constraining functions that cohere with the specifications that quantum theory provides individually for spin-product results of the function domain settings. To produce such an assessment is the burden of our final computations. This involves investigating the implications of Bruno de Finetti's "fundamental theorem of probability" for the nine proposed Stern-Gerlach gedankenexperiments, all performed on the same pair of electrons.

Bell's quantum probability polytope computed via de Finetti's FTP

If you are not familiar with Bruno de Finetti's (1974, 3.10) fundamental theorem of probability, the FTP, please review its brief description and the commentary already presented in Chapter 1. In a word, the theorem identifies a linear programming problem that characterises the bounds on the probability for any event that are required by its coherency with other probabilities that are taken as given. *Presuming your basic familiarity*, the remainder of this Section presents an analysis of the implications of theoretical quantum probabilities for the results of the now classic gedankenexperiment on a pair of electrons, propelled toward the Stern-Gerlach magnets of Alice and Bob. It covers a description and a formalisation of QM-motivated probability assertions appropriate to the several observations considered, and displays computations of the bounding implications for other probabilities about which the theory is silent. Although quantum theory says nothing precise about the gedanken results, it places definitive restrictions on what would cohere with what it does say.

Our first exposition is merely discursive, describing conversationally the setup of the linear programming problems relevant to our discussion. It does not dwell on explicit formal definitions of all notation, but rather proceeds straightforwardly with discussion using vocabulary that is standard in LP methods. It is followed by presentation of formal algebraic detail describing the quantities and constraints involved in these LP problems. Numerical results then portray the polytope of probability vectors that represent the coherent content of recognised quantum theory.

LP problem identifying quantum probabilities that recognise restricted non-linear functionals

To begin, we shall presume standard quantum expectations and probabilities for spin-products in the domain experiments 23, and use linear programming to specify bounds on implied probabilities for spin-product observations in the range experiments 1456789 imagined to be concurrent in the thought experiment. Once we are clear on how this works, we shall describe how such a set of min/max computations would be replicated for other function domains, and then concatenated for all twelve functional restriction structures.

Familiar by now with the quantum probability distributions for the outcomes of any real experiment, recall that the probabilities for the four possible outcomes of the spin observations, $++, +-, -+,$ and $--$, can all be identified from the probability for any one of them, say P_{++}. For the four quantum probabilities resolve to $P_{++} = P_{--}$ along with $P_{+-} = P_{-+} = \frac{1}{2}[1 - 2P_{++}]$. Furthermore, this probability is uniquely related to the expected value of the spin-product via the equation $E(AB) = P_{++} - P_{+-} - P_{-+} + P_{--}$, which then equals $4P_{++} - 1$. The probability for observing a spin-product value of -1 resolves to $P[AB = -1] = [1 - E(AB)]/2$.

In the context of any component experiment for which the relative angle between the two magnet orientations is equal to θ, these relations specify

$$P_{++}(\theta) = \tfrac{1}{2}[1 - cos^2(\theta/2)], \text{ and } E[AB(\theta)] = 1 - 2cos^2(\theta/2).$$

For reference in our computations relevant to Mermin's problem, these spin-product expectations resolve to

$$E[AB(-120°)] = E[AB(+120°)] = .5, \text{ and } E[AB(0°)] = -1.$$

Correspondingly,

$$P[AB(-120°) = -1] = .25, \text{ and } P[AB(0°) = -1] = 1.$$

Here is the problem, stated directly in conversational English, presuming familiarity with all algebraic notation and detail. We are to investigate the QM-motivated probability specifications for the four possible observation vectors of nine spin-products observed by Alice and Bob in a 3×3 paired-angle-thought-experiment on the same pair of electrons. Each of these vectors specifies an array of values for all nine components of the spin-product gedanken vector we designate by

$$\mathbf{G_9} \equiv (A_nB_n, \ A_nB_z, \ A_nB_p, \ A_zB_n, \ A_zB_z, \ A_zB_p, \ A_pB_n, \ A_pB_z, \ A_pB_p)^T.$$

These are the quantities that are crucial to the specifications of quantum theory as it pertains to Bell-type considerations. A complete list of their possible gedanken observations is constituted by the four columns of the left half of the middle partition of the realm matrix we created for all quantities involved in the problem (the six electron spins and their nine paired products). We designated this sub-matrix as $\mathbf{R_{9,4}}$. Only four of the nine rows of this matrix are distinct, the other five being repetitions. Refresh your memory by examining it again within the full matrix, shown on page 115.

Among these possible 9-dimensional vectors of spin-products, there are only two dimensions of free observations, on account of the functional restrictions embedded within them. For example, the observations at the magnet configurations numbered 2, 3 would functionally identify the results at configurations 1, 4, 5, 6, 7, 8, 9 according to quantum theoretic specifications, enhanced by Einstein's presumed principle of local realism. However, this functional relation is evidently non-linear, for the rank of the realm matrix of nine spin-product possibilities is 4. This rank corresponds to any four distinct rows of $\mathbf{R_{9,4}}$. Thus, the specification of quantum expectations pertinent to the domain configurations of rows 2 and 3 would place only polytopic bounds on the cohering probabilities for the other spin-products they imply.

If we were to specify a *complete* probability distribution vector, q_4, over the possible spin-product outcome components of the function domain, $\{-1, +1\}^2 = \{(-1, -1), (-1, +1), (+1, -1), (+1, +1)\}$, our problem would be over. However, quantum theory *explicitly disavows* an assertion of a *complete* distribution vector q_4 over these possibilities. For this would entail a specification of joint probabilities for the results of non-commuting measurement operators on the state space of the electron pair. While quantum theory specifies precise probabilities for the two possible values of the spin-product observation $A_{d_A} B_{d_B}$ at any experimental paired angle setting of Stern-Gerlach magnets, it explicitly says nothing about the *joint outcomes* of the spin-products observed at *both* settings 2 *and* 3, for example. Nonetheless, quantum theory *does* specify explicitly cohering probabilities regarding the outcomes of the spin-product experiment at each of the configurations 2 and 3 separately. Observations at this pair of settings constitute the domain of the function we have designated as $23 \to 1456789$.

An aside of detail should clarify the preceding disavowal. While QM theory would clearly specify joint probabilities such as

$$P[(A_n = 1)(B_z = -1)] \quad \text{and} \quad P[(A_n = 1)(B_p = +1)]$$

for example, deriving from the expectations $E(A_n B_z)$ and $E(A_n B_p)$, it explicitly disavows more detailed assertions of the form

$$P[(A_n = 1)(B_z = -1)(A_n = 1)(B_p = -1)].$$

The former two assertions each specify a standard specification of quantum theory relative to a pair of spin observations at a selected paired angle setting; whereas the latter assertion would entail a claim about joint observations of spins involving both B_z *and* B_p, observations represented by non-commuting measurement operators. However, a joint probability of this sort would be required in order to specify precise values for components of the vector q_4, flouting this abstemious honesty. Quantum theory does *not* provide for a complete distribution of probabilities for the four possible spin-product components of $\{-1, +1\}^2$, neither at the domain pairing 23, nor any other pair of domain variables among the twelve functions embedded in the realm of possibility for the nine gedanken products.

Suppose that we entertain the cohering expectations of quantum theory pertinent to the isolated domain products $G_2 = A_n B_z$ and

$G_3 = A_n B_p$, viz., $E(A_n B_z) = 1/2$, and $E(A_n B_p) = -1/2$. What would these imply for the cohering expectations of spin-products $1, 4, 5, 6, 7, 8$, and 9 determined in their function range? Based on the associated rows of the realm matrix for the spin-products, these expectations would place two linear restrictions on the components of any prospective gedanken vector $\mathbf{q}_4$. Along with the constraint that their components are all non-negative and sum to 1 (unity), a linear programming routine would identify a pair of solution vectors $\mathbf{q}_{4min}$ and $\mathbf{q}_{4max}$ that produce the extreme feasible values of the objective functions $E(A_d B_d)$ for any range orientation pairing, subject to the quantum theoretical linear constraints on spin-products 2 and 3. (These are in addition to the requirement that $E(AB) = -1$ for the spin-products of orientation pairs 1, 5, and 9. This is the condition specified by quantum theory that underlies the entire problem, and which restricted the realm matrix.) The linear coefficients of the objective functions can be identified from appropriate rows of the realm matrix.

As it turns out, with expectations specified for any pair of two domain variables, there is only one range variable whose extreme cohering expectations we shall need to investigate. Remember that the principle of local realism implies the equality of the spin-products $A_n B_z = A_z B_n$, $A_n B_p = A_p B_n$, and $A_z B_p = A_p B_z$. These correspond to the identity of rows 2 and 4, 3 and 7, and 6 and 8 in the realm matrix $\mathbf{R}_{9,4}$. Thus, with expectations settled at -1 for the negative spin-products at orientations 1, 5, and 9, asserting quantum theoretic expectations at orientations 2 and 3 would imply the same expectations at orientations 4 and 7 as well. *This would leave only spin-product expectations for orientations 6 and 8 to be investigated, and these must be identical.* Thus, a single pair of min/max linear programming problems would identify the bounds on the entire cohering expectation vector $E(\mathbf{G}_9)$. We shall specify the algebraic details of this computational routine now.

The paired solution vectors $\mathbf{q}_{4min}$ and $\mathbf{q}_{4max}$ contain the information that puts bounds on the general problem solution we seek: to identify the extreme vectors $\mathbf{q}_4$ that both satisfy *all* quantum probability specifications relevant to the gedankenexperiment, and support appropriate cohering range expectations. We will need to determine this pair of solution vectors for the min/max LP problems appropriate to each of the twelve function domains we have identified. These would enclose the entire space of $\mathbf{q}_4$ vectors that represents the implications of quantum theory pertinent to the magnetic spin-product gedankenexperiment on an electron pair.

Before presenting the numerical results of these computations, let's state concisely the linear restrictions of the several linear programming problems we have delineated.

Algebraic representation of the LP constraints

Our goal is identify the prognostications of quantum theory regarding a gedankenexperiment: Mermin's physics problem of two electrons engaging the nine 3×3 design of Stern-Gerlach magnets at the stations of Alice and Bob.

Let $\mathbf{G}_9$ denote the ordered column vector of imagined spin-product outcomes of the nine experiments on a single pair of electrons:

$$\mathbf{G}_9 \equiv (A_nB_n,\ A_nB_z,\ A_nB_p,\ A_zB_n,\ A_zB_z,\ A_zB_p,\ A_pB_n,\ A_pB_z,\ A_pB_p)^T.$$

Let $\mathbf{R}_{9,4} \equiv \mathbf{R}(\mathbf{G}_9)$ designate the realm matrix of possibilities for this gedankenobservation vector. We have already displayed it as the left half of the middle matrix partition of the large realm matrix you have already reviewed. We shall refer to individual *rows* of this matrix using the denotation $\mathbf{r}_{i\cdot}$ for values of the row numbers $i = 1, ..., 9$, and to individual *columns* of this matrix by $\mathbf{r}_{\cdot j}$ for values of the column numbers $j = 1, 2, 3, 4$.

Let $\mathbf{1}_4$ denotes a *row* vector of four 1's.

Let $\mathbf{b}_9$ denote the *column* vector of numerical values of quantum theoretic expectations for the components of $\mathbf{G}_9$ when they each designate the outcome of a single real experiment on a pair of electrons at a specific angle pairing of the S-G magnets.

Finally, let $\mathbf{Q}_4$ denote the column vector of events whose components identify whether the observation vector $\mathbf{G}_9$ would happen to equal the various columns of $\mathbf{R}_{9,4}$. That is, the component $Q_j = (\mathbf{G}_9 = \mathbf{r}_{\cdot j})$. Each of these four events equals 1 or 0, and only one of them equals 1.

Notice that whereas quantum theory specifies expectation values for the components of $\mathbf{G}_9$ when each is entertained as the spin-product of a unique pair of electrons in an individual experiment on its own, it does not specify expectations for the components of $\mathbf{Q}_4$. For these events identify the joint outcomes of several incompatible experimental observations. However, the linear programming problems we now address do codify restrictions on the space of such expectation vectors that would cohere with what quantum theory specifies about the domain experiments individually.

To begin, we formalise the linear programming investigations required by the spin-product function $23 \to 1456789$ yielding bounds on $E(G_6)$, as we have discussed informally in the details of the preceding subsection:

Find the vectors $\mathbf{q}_{4min}$ and $\mathbf{q}_{4max}$ that yield

the minimum and the maximum values of $\mathbf{r}_6 \cdot \mathbf{q}_4$

subject to the conditions that

$$
\begin{pmatrix} \mathbf{r}_{2\cdot} \\ \mathbf{r}_{3\cdot} \\ \mathbf{1}_4 \end{pmatrix} \mathbf{q}_4 = \begin{pmatrix} b_2 \\ b_3 \\ 1 \end{pmatrix} \quad ,
$$

where each component $q_i \geq 0$.

We shall denote these two solution vectors by $\mathbf{q}_{4min23\cdot6}$ and $\mathbf{q}_{4max23\cdot6}$. For once we determine them, we shall need to repeat such computational LP searches so as to determine extreme vectors appropriate to the other eleven spin-product functions that govern possibilities for a different single component of the spin-product vector $\mathbf{G}_9$. Formally, this would amount merely to changing the coefficient vectors $\mathbf{r}_{2\cdot}$ and $\mathbf{r}_{3\cdot}$ in the LP domain constraints, and changing the objective function coefficients $\mathbf{r}_{6\cdot}$ accordingly so to represent the range variable whose expectation bounds we search. In principle, there could be 24 such extreme solution vectors. However, on account of duplications among the row vectors of $\mathbf{R}_{9,4}$, the number of distinct problems reduces to merely three distinct LP problem pairs, and six distinct solution vectors. It should be evident, for example, that $\mathbf{q}_{4min23\cdot6} = \mathbf{q}_{4min47\cdot8}$ and $\mathbf{q}_{4max23\cdot6} = \mathbf{q}_{4max47\cdot8}$. The LP problem that yields either one of these solutions is numerically identical to that which yields the other. For rows 2, 3, 6 of the realm matrix $\mathbf{R}_{4,9}$ are identical to rows 4, 7, 8.

It is a feature of coherent probability structures that any convex combination of coherent expectation assertions is also coherent. It is the convex hull of all the six extreme expectation vectors that represents the quantum theoretic prognostications for the gedanken-experiment. We shall now display the six vectors.

Computational results

The six solution vectors that resolve the three distinct pairs of LP problems involved in this investigation are displayed in the top section of Table 3. Notational subscripts on the solution vectors specify the form of the LP problems from which each derived. Although they were designed to identify extreme values of expected spin-products at specific magnet orientations, the probabilities underlying each of these extreme solution vectors would specify cohering expectation values for every one of the spin-products that result from a "run" of the gedankenexperiment at all nine relative angle configurations. These implied nine-vectors of expected spin-products are printed in Table 3 immediately below the solution vectors to which they correspond. The rank of the matrix of solution vector columns is 4.

Table 3. Extreme spin-product expectations for unique LP solutions

LP solutions	q_{4min} 23·6	q_{4min} 26·3	q_{4min} 36·2	q_{4max} 23·6	q_{4max} 26·3	q_{4max} 36·2	QM	Sim
q_1	.25	.25	.25	0	0	0		
q_2	0	0	.75	.25	.25	.5		
q_3	0	.75	0	.25	.5	.25		
q_4	.75	0	0	.5	.25	.25		
E(spinprod)							QM	Sim
$E(A_nB_n)_1$	-1	-1	-1	-1	-1	-1	-1	-1
$E(A_nB_z)_2$	.5	.5	-1	.5	.5	0	.5	.25
$E(A_nB_p)_3$	.5	-1	.5	.5	0	.5	.5	.25
$E(A_zB_n)_4$	.5	.5	-1	.5	.5	0	.5	.25
$E(A_zB_z)_5$	-1	-1	-1	-1	-1	-1	-1	-1
$E(A_pB_z)_6$	-1	.5	.5	0	.5	.5	.5	.25
$E(A_zB_p)_7$	.5	-1	.5	.5	0	.5	.5	.25
$E(A_pB_n)_8$	-1	.5	.5	0	.5	.5	.5	.25
$E(A_pB_p)_9$	-1	-1	-1	-1	-1	-1	-1	-1

To the right of the six E(spinprod) vectors in Table 3 appear two additional column vectors for purposes of comparison. The column headed by **QM** exhibits the standard expectations of quantum mechanics for the spin-product observed in an actual experiment at any *single one* of the various paired angle configurations. The final additional column headed by **Sim** exhibits expectations corresponding to the proportions of differing spin observations $(-+)$ or $(+-)$ that were generated in our simulated gedankenexperiments on magnetic spins, rounded to .375. The simulations, remember, relied upon quantum theoretic probabilities to generate the emergence of spin-products at each of the domain configurations, and then relied on the functional relations to yield the other seven dimensional components accordingly. The simulated behaviour of each electron pair involved its engaging the magnet orientations at detection stations in all nine of their experimental paired orientations. This is the situation that provoked Mermin's rejection of the message-encoded balls explanation.

The glory of Table 3 is that its column q_4 vectors specify the extreme points of a bounding polytope of expectation vectors supported by quantum theory, as it would be relevant to the spin-products of a gedankenexperiment on a single pair of electrons at all nine of the possible orientation pairings. The fundamental theorem of probability does not identify a specific vector of expectations for spin-products at every one of the nine paired magnet orientations. Neither do the prescriptions of quantum theory, which avoid precise assertions regarding the *joint* outcomes of non-commuting measurements. Rather, they identify the space of allowable expectation vectors that would cohere with the assertions about real experiments that quantum theory actually is endowed to assess. Based on the quantum probabilities that constrained the several LP computations, the conclusion to this exercise is that the sought-for vector of gedanken expectations needs merely sit somewhere within the convex hull of the E(spinprod) vectors that appear in the first six columns of the second bank of Table 3.

The rank of the matrix of 9-D expectation vectors appearing in Table 3 is only four. These four dimensions are spanned by the rows $1, 2, 3,$ and 6 of the Table. The other rows are repeats of these, so we could have listed four different row numbers, but the result would be the same. In order to discuss these results in the same terms with which Professor Mermin assessed the behaviour of his machine, we shall transform the expectations shown in Table 3 into the probabilities they imply for negative spin values, displaying these in Table 4. This linear transform is $P[AB(\theta) = -1] = \{1 - E[AB(\theta)]\}/2$ for each relative angle between the magnet orientations.

Table 4. Probabilities for negative spin-products along extreme solution vectors

LP solutions	q_{4min} $23 \cdot 6$	q_{4min} $26 \cdot 3$	q_{4min} $36 \cdot 2$	q_{4max} $23 \cdot 6$	q_{4max} $26 \cdot 3$	q_{4max} $36 \cdot 2$	QM	Sim
$P(A_n B_n = -1)_1$	1	1	1	1	1	1	1	1
$P(A_n B_z = -1)_2$	.25	.25	1	.25	.25	.5	.25	.375
$P(A_n B_p = -1)_3$	.25	1	.25	.25	.5	.25	.25	.375
$P(A_z B_p = -1)_6$	1	.25	.25	.5	.25	.25	.25	.375

One feature of these computations is that we can now make sense of the simulation results we generated in Table 2. These defied Professor Mermin's claims regarding what would obtain in such (impossible) experimentation on a single electron pair. The simulated counts of negative spin-products (matching light signals) displayed there yielded proportions on the order of .375 for six of the

Stern-Gerlach orientation pairings. These differ markedly from the Mermin proclamation that proportions of negative spin-products would hardly differ from .25 at these settings. In fact, they exhibit the order of magnitude that he found upsetting, exceeding $1/3$.

Recognising the first six columns of Table 4 as vertices of a quantum theoretical probability polytope for negative spin-products, we can determine that the appended final column vector headed by **Sim** *is located within* their convex hull. Algebraically, it is equal to the convex combination of those columns with convexity coefficients $(1/6, 0, 1/12, 0, 1/2, 1/4)$. In contrast, the appended column headed by **QM** does *not* lie in this convex hull. It cannot be expressed as a convex combination of the vertex vectors. To proclaim it as representing the prescriptions of quantum mechanics for the outcome of the gedankenexperiment would be incoherent. The accompanying relegation of Einstein's principle of local realism amounts to nonsense. We can view the situation geometrically in Figure 2.

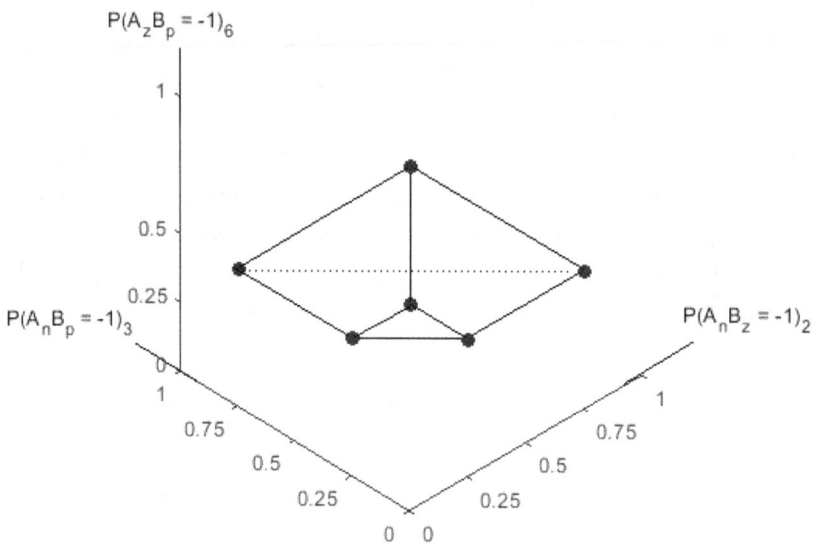

Figure 5: A three-dimensional polygon constituting the convex hull of quantum probabilities for negative spin-products in the gedankenexperiment when the Stern-Gerlach magnet directions are not set identically.

Since the top row of Table 4 is constant at the value of 1, we can recognise that a convex hull enclosing the column vectors of this matrix constitutes a 3-D polygon within a hyperplane in the

4-D space. It appears in Figure 2 as a tetrahedron that has lost one of its tips. Professor Mermin's proclaimed point of probabilities in these three dimensions, $(.25, .25, .25)$, is exterior to this polytope, while the simulation vector of probabilities $(.375, .375, .375)$ is a point well within the hull as a convex combination of its vertices. When the operation of Professor Mermin's machine is applied to the gedankenexperiment, the crude vector of quantum probabilities representing his provocative claims lies outside of the convex hull of probability vectors that are supported by the results of quantum theory. The professor's vector would constitute the front tip of the tetrahedron, but the constraints that are commonly ignored in the problem have cut off this tip. His assertions and his concerns derive from a mathematical error of neglect, similar in structure to the error we have found in the claims of Aspect/Bell.

Behind the smoke and mirrors: a mystery exposed

In challenging our magician, we have now been through an exercise of tedious mathematics and computation. It is time to conclude with an overview of what we have learned. I will write quite frankly, leaving a relevant discussion of the foundations of probability to Chapter 6.

There is nothing wrong at all with Professor Mermin's machine, and furthermore there is nothing mysterious about it. The machine quite accurately portrays the probabilistic structure of the current propositions of quantum theory regarding the experimental observation of magnetic spins of paired electrons as they pass angled Stern-Gerlach magnets at two distantly separated detection stations. To be explicitly precise, the machine is designed to exemplify the structure of theoretical and empirical results from sequential observations of distinct electron pairs ejected to any one of nine different paired experimental settings, similar in design. Each of the paired dial settings on the two machines constitutes a different type of experiment. At three of the paired angle settings when the two anglings are identical, every observation of the spin-product equals -1. The spins (A, B) observed at the two stations are recorded as either $(-1, +1)$ or $(+1, -1)$. At any one of the six other paired angle settings, long sequences of observations yield spin-products equal to -1 in $1/4$ of the runs, and $+1$ in $3/4$ of the runs. For any individual unique experiment, quantum theory asserts only the probability of $1/4$ for the negative spin-product.

Many regard this result to be mysterious in itself, involving what Einstein had referred to as "spooky action at a distance". The source of this attitude derives from imagining that the probabilistic behaviour of the results correspond to a feature of randomness inherent in the particle activity itself. Nature at its fundamental base is considered to be random, governed by recognisable probability distributions, derivable by theoretical acumen. It is this conception of the matter to which Einstein objected. He proposed rather, with EPR, that the stochastic aspect of quantum theoretic results derives from the incompleteness of the theory and from our uncertainty regarding the influence of unknown and unobserved "supplementary variables" pertinent to the exact conditions of any specific experimental run. Heretofore unobserved by us, these would presumably vary from run to run. This proposal would relegate "the mystery" of quantum results to the same category of mystery involved in common mysteries of activity at classical scales of magnitude, such as whether today's milk yields from two specific dairy cows exceed fifty kilograms today. Maybe (yes, yes), maybe (yes, no), maybe (no, yes), maybe (no, no). As with the milk yields, it is the codification of symmetries in our uncertainty regarding conditions of the experimental quantum problem that yield the probabilistic specification of the mechanics.

When regarded as properties of the particles themselves, the quantum probabilities of Mermin's machine do seem to pose the mysterious question of how the probabilistic activity of the ball at station A can depend on the dial (magnet angle) setting at station B if there is no way for the status of the dial at B to be communicated to station A when a ball (an electron) arrives there. I will devote Chapter 6 of this book to an explication of how the understanding of probabilities as measures of our uncertainty about unknown quantities can clearly resolve such mysteries. But for now, we shall focus the conclusion of this chapter on what the professor proclaims, erroneously, as an even deeper mystery. It concerns his attitude, and the virtually universal attitude of quantum theorists today, that the defiance of Bell's inequality in a gedankenexperiment on a single pair of electrons at all nine experimental settings defies the proposition of local realism, supplementary variables, and the uncertainty interpretation of the physical situation in accounting for the evidence of quantum experiments.

Professor Mermin models the action of such supplementary variables by encoded designations of colour schemes imprinted on the balls it ejects toward the stations. He proposes this as a model of any supplementary variables explanation of the quantum experiment, on the same metaphorical order as that of the machine with balls modelling the observation of electron spin behaviour. Such a proposal could be fair enough, although one must surely quibble with its adequacy for representing the substance of the supplementary variables viewpoint. Nonetheless, his analysis of the machine's response to an encoded pair of balls when they arrive at all nine dial settings generates what he considers to be an even deeper mystery. Although pleased that the scheme ensures matching light signals when the dials are set identically, he bristles at the fact that the proportion of matching lights (spin-products equal to -1) deriving from such a scheme will exceed $1/3$. It is this result, which he considers to mirror the supposed defiance of Bell's inequality, that is seen to constitute a defiant challenge: that no other supplementary features of the experimental situation can account for the known behaviour of quantum experiments. Spin-products are observed to equal -1 in only $1/4$ of the experimental runs on electron pairs.

It is this result that is just plain wrong. Examining the real quantum experiment which the professor would have us ignore, we find embedded within the corresponding thought experiment a surprising feature that has long been unnoticed. Subjecting each pair of electrons to spin-detection at all nine of the paired angle settings would engender an array of restrictive functional relations among the nine observed spin results. The professor neglects these symmetric functional relations mapping $\{-1, +1\}^2 \rightarrow \{-1, +1\}^7$ in his analysis of the gedankenexperiment. His claims regarding quantum theoretic prescription of matching lights in $1/4$ of such gedanken observations when the switches differ are blatantly false. They rely on the possibility that any string of nine-tuples deriving from the Cartesian product $\{-1, +1\}^6$ of possibilities could designate the outcome of such a thought experiment. We have seen otherwise ... that many such strings are impossible. The space of possibilities derives rather from the structure $\{-1, +1\}^2$, replicated in the different pairs of angled magnet orientations. Moreover, we have used computational procedures of linear algebra to identify precisely the polytope of probabilistic assertions regarding the outcomes of a gedankenexperiment that represent the honest claims of quantum theory regarding this matter.

The probabilities for matching lights proclaimed by Mermin are representable by a nine-tuple vector that does not lie within the convex hull of the coherent vectors supported by quantum theory. We have created a Monte Carlo simulation of results of a scenario that is both wholly consistent with this theory and also respects the restrictive symmetric functional relations that derive from the structure of the experiment. It generates proportions of matching lights on the order of .375, precisely on the order of magnitude that the professor would have us suspect on account of his shenanigans. This vector does reside within the convex hull required by coherency. But there is still more to say about this!

One of the most famous features of quantum theory, known widely by name to the general public, is the relevance of Heisenberg's uncertainty principle. Under the guise of that name, the principle concerns physical experiments with electrons that attempt to measure both the position and the velocity of the electron at a point in time. What the principle says is that actually it is impossible to make such a joint measurement of both of these characteristics of the electron at the same time. We can make a measurement of one or the other, but not both. Technically, the quantum theory identifies this impossibility by the characterisation of the two measurements of the quantum state via measurement operators that do not commute. Quantum theory can specify probability distributions for possible values of either of these measurements on an electron. However, it cannot and does not provide an assessment of a joint probability for the observation of both measurements. For such an operation is impossible. The incommensurability of measurements is characterised theoretically by matrix operators on states that do not commute. This is the basis for the general form of the uncertainty principle of theoretical quantum mechanics.

The incommensurability of simultaneous observation of paired electron spins at several different paired magnet orientations in a gedankenexperiment belies the professor's claims about his machine in this context. Quantum theory does not identify a joint probability distribution for the results of all nine of them. It does provide precise probabilities for the four possible observations of spin pairs at any single paired orientation, these being $++, +-, -+,$ and $--$

at the two detection stations. Moreover, it can stipulate probabilities for such outcomes from several distinct experiments on an electron pair at differing magnet angles, realising albeit that only one of them can be engaged. These are the probabilities used in linear programming routines that identify the restrictions on range probabilities provided by individual assessments of domain probabilities. However, it leaves four dimensions of freedom remaining unascertained for the joint distribution of all nine measurements. The probability distributions of quantum theory for the results of a real experiment on an electron pair at any one of the nine design orientations *may not* be considered to be a marginal distribution from the joint distribution. There is no joint distribution over these imagined experimental results supported by quantum theory! Any vector within the polytope of feasible joint distributions can be transformed mechanically into a vector of marginal probabilities for the spin-products at any magnet orientation pair, but no one of these constitutes a distribution that is marginal with respect to a joint distribution over all feasibilities.

In particular, the result of our simulated experiment, using the probabilistic assertions of quantum theory relevant to the domains of the restrictive functional relations, cannot be presumed to suggest a definitive proclamation of quantum mechanics. Its generation had embedded into it a formulation of independence among the outcomes of any two spin-products observed among the domain arguments of the functions. While feasible in the context of the agnostic stance of the theory relative to incommensurable measurements, this is not a requirement of the theory. Furthermore, in the context of the active claims of the theory regarding the entanglement of spin observations at distant detection stations, it might well be suspect by those who might like to think about such things. We are left with the realisation that quantum theory provides only a convex polytopic boundary of probabilities for the result of the gedankenexperiment to which Bell's inequality is relevant. For none of the distributions within the polytope that is entertained is the inequality defied.

Appendix to Chapter 4:
on deeper functional relations

While investigating the realm matrix of possible observation values for the vector $\mathbf{G}_9$ in a gedankenexperiment, we found firstly that the 512 apparent possibilities for the spin-product vector of the were reduced to at most 64 by the principle of local realism. These were then reduced still further to only 8 by the specific definitive prognostication of quantum theory when the magnet orientations at the detector stations are identical: $P(A_d B_d = -1 | \theta = 0) = 1$, and equivalently $E[A_d B_d | (\theta = 0)] = -1$. Whenever the two magnet orientations are the same, whatever their shared directions relative to vertical, then the observed value of the spin-product is surely equal to -1. We might also remember that the original argument of EPR concerning the incompleteness of quantum theory rested on a scenario in which the quantum probabilities are similarly definitive. It was only under conditions posited by Bell that experimentation was considered under more liberal conditions. In Bell's scenario, Alice and Bob might each select a detector orientation from a pair for which both differ from the pair available to the other. This led to the four relative angles between the polarisers in the CHSH formulation of Bell's inequality: $\theta = -\pi/8, -3\pi/8, \pi/8$, and $-\pi/8$. In none of these conditions is the prognostication of quantum theory definitive. In such a scenario was found the apparent violation of Bell's inequality that we debunked. The standard mistaken analysis neglects the functional relations inherent among the possible polarisation products of the detections at the two stations in a gedankenexperiment on the same pair of particles.

I should like only to remark here that if the scenario of Mermin's quantum experiment were expanded so as to allow the three magnet orientations at stations A and B to be completely distinct (denoted perhaps as $\mathbf{a}, \mathbf{a}', \mathbf{a}'', \mathbf{b}, \mathbf{b}'$, and $\mathbf{b}''$), then the relative angles between their paired orientations would never equal 0, and the prescription of quantum theory about the spin-product would never be definitive. In such a scenario, the 64 possible spin-product vectors allowed by the principle of local realism would not reduce to merely 8 by reasons of quantum probabilities, and then to 4 by functional relations inhering among the remaining possibilities. However, among the full gamut of 64 possibilities allowed by local realism in such a scenario, it turns out that there are embedded even more ornate functional relations among them: again, these amount to twelve such functions, but each of them maps $\{-1, +1\}^5 \to \{-1, +1\}^4$. These reduce the number of conceivable outcomes for the spin-product vector from 64 to only 32. I will merely list the embedded functions here for your edification, and present the analysis identifying them elsewhere:

$12359 \to 4678$	$13459 \to 2678$	$14579 \to 2368$
$12569 \to 3478$	$13569 \to 2478$	$14589 \to 2367$
$12579 \to 3468$	$13589 \to 2467$	$15679 \to 2348$
$12589 \to 3467$	$14569 \to 2378$	$15789 \to 2346$

If you pull the components $1, 5, 9$ out of these domain identifiers and place them among the range identifiers, you will recognise the twelve functions that we have determined are appropriate to Mermin's more restricted problem. The conclusion of a complete analysis of the less restrictive scenario is that a proposition of Bell-inequality-defying-probabilities is precluded by recognition of these functional relations. The proposition could be considered to be challenging if and only if these functional relations were mistakenly ignored.

Chapter 5

SIMULATING
MULTI-POLARISATION
IN A GEDANKENEXPERIMENT:
the irrelevance of empirical averages

Bell's inequality in CHSH form pertains to a thought experiment that cannot actually be conducted. It is imagined to yield a linear combination of four polarisation products observed simultaneously on a single pair of photons. It is universally agreed that if the principle of local realism holds for quantum phenomena, then the imagined observable values of this combination are limited to $\{-2, +2\}$. This limitation constitutes the basis for Bell's inequality. If four *distinct* photon pairs were used in its component experiments, the possibilities for the combination $s(\lambda)$ would extend from $\{-2, +2\}$ to $\{-4, -2, 0, +2, +4\}$, allowing any cohering expectation value within the interval $[-4, +4]$ rather than merely $[-2, +2]$. An evidential evaluation of quantum defiance of the inequality requires a properly constructed simulation routine that complies both with the prescriptions of quantum theory and with the principle of local realism. I provide one and compile it in this chapter.

Widely respected argument to the contrary portrays empirical confirmation of the defiance as deriving from a sampling theoretic characterisation of experimental observations. Leading exposition of this outlook by Gill (2014) has formalised a line of argument running through the reports of the Aspect group (1981, 1982) as motivated in his Bell memorial lecture (Aspect, 2002), and subsequently in works of Redhead (1987), Gill (2003), and Mermin (2005). These suggest that sequences of observations of distinct photon pairs at single randomly selected polarisation angles can be considered to be sample observations of simultaneous polarisation products for the same pair of photons at an array of four angles. Coupled with a common misconception that expectations are averages, the reigning understanding in the physics community is that the defiance of Bell's inequality by quantum probabilities is a matter of experimental fact. The eminent journal *Nature* has boldly pronounced the "death by experiment for local realism" (Wiseman, 2015).

In this chapter we shall display the results of a simulation study of the Aspect-Bell experiment that defies such a characterisation of the situation, supporting rather the challenging analysis presented here in Chapter 1, and originally published in Lad (2021). This shows that quantum theory does not provide an expectation of the polarisation combination devised by Clauser, Horne, Shimony, and Holt (1969) as $2\sqrt{2}$. Rather, it supports only an interval expectation of $(1.1213, 2]$.

I shall not rehash once again the well-known activities of Alice and Bob in observing the paired photons' behaviour, described in Chapter 1. Rather I shall turn immediately to describe the simulation procedures I assess. Discussion of the results and relevant matters will follow. Suffice it to say only that I am following Gill's simplified notation (2014) for the polarisation observations A, B, A', and B'. These occur at four relative polariser angles, designated by Aspect as $(\mathbf{a}, \mathbf{b}) = -\pi/8$, $(\mathbf{a}, \mathbf{b'}) = -3\pi/8$, $(\mathbf{a'}, \mathbf{b}) = \pi/8$, and $(\mathbf{a'}, \mathbf{b'}) = -\pi/8$. Each of the observable four polarisation products $AB, AB', A'B$, and $A'B'$ can equal only -1 or $+1$. For any angle pairing, quantum theory specifies the probabilities

$$P[A(\mathbf{a}^*) = +1 \text{ and } B(\mathbf{b}^*) = +1] = \tfrac{1}{2}cos^2(\mathbf{a}^*, \mathbf{b}^*)$$
$$= P[A(\mathbf{a}^*) = -1 \text{ and } B(\mathbf{b}^*) = -1],$$

where $(\mathbf{a}^*, \mathbf{b}^*)$ denotes the relative angle between any two polariser directions $\mathbf{a}^*$ and $\mathbf{b}^*$ at the observation stations. The directions are equally likely to be selected in a specific experiment as $\mathbf{a}$ or $\mathbf{a'}$, and as $\mathbf{b}$ or $\mathbf{b'}$, respectively. The companion quantum probabilities are the exchangeable

$$P[A(\mathbf{a}^*) = +1 \text{ and } B(\mathbf{b}^*) = -1] = \tfrac{1}{2}sin^2(\mathbf{a}^*, \mathbf{b}^*)$$
$$= P[A(\mathbf{a}^*) = -1 \text{ and } B(\mathbf{b}^*) = +1].$$

The simplified notation assigns $A = A(\mathbf{a}), B = B(\mathbf{b}), A' = A(\mathbf{a'})$, and $B' = B(\mathbf{b'})$. In the context of Aspect's angle specifications, the quantum theoretic expectations of the polarisation products are

$E(AB) = E(A'B) = E(A'B') = 1/\sqrt{2}$, while $E(AB') = -1/\sqrt{2}$. If experiments were to be conducted on four distinct pairs of photons, each following its own generation procedure, the observable value of $s \equiv AB - AB' + AB + A'B'$, a combination of products, might equal any of $\{-4, -2, 0, +2, +4\}$, with quantum expectation indisputably $E(s) = 2\sqrt{2}$.

In contrast, if the experiments were imagined to be conducted impossibly on the same pair of photons to assess their *joint* state with respect to all four polarising angles and attendant conditions, then the value of s could equal only one of the values $\{-2, +2\}$, agreeing with Bell. Such a restriction would derive from presuming that the principle of "local realism" would govern quantum phenomena, just as it is presumed to govern physical phenomena at classical scales of activity. Local realism would entail that the value of A exhibited in an experiment conjoint with the observation of B, say, would need be identical to its value in a concomitant experiment conjoint with observation of B'. For the local conditions surrounding A would be the same both when the paired polariser were set so as to observe either B or B'.

Of course, in a real experiment we can only observe the polarisation product for a photon pair at *one* of the four relative polariser angles. However, the restriction of the quantity s to equal either -2 or $+2$ would pertain only to a combination of imagined observations from experiments conducted at all four angles with the same pair of photons. In tandem with this restriction arises an array of linear functional relationships engendered among the polarisation products in the imagined experiment, which do *not* govern relations among products in actual experiments on four distinct photon pairs. As we shall observe in the simulations we now create, it is the neglect of these four functional relations among the constituents of Bell's quantity in the CHSH format that underlies the mistaken understanding that quantum probabilities defy his inequality on the basis of observable averages.

It needs to be said that quantum theory itself says nothing about such speculations. The uncertainty principle precludes the prescription of properties to experimental results that cannot possibly be observed. Nonetheless, it is just such suppositions that underlie the long-running controversy regarding Bell's inequality and its relevance to quantum theoretic probabilities. Let us turn directly to the simulation, and defer discussion until the results are at hand.

Setting up the simulation

The simulation involves a sampling of tendered observations on the components of s. These would be polarisation products derived from imagined sequences of joint observations on all four values of A, B, A', and B' generated by single pairs of photons. As does Gill (2014), we shall consider a spreadsheet containing an $N \times 4$ table of products that derive from these, each equal to either -1 or $+1$. The rows will be labelled by an index $j = 1, ..., N$, while the columns are labelled with names of the four polarisation products for a photon pair. Thus, the jth row of the spreadsheet is composed of the imagined products $A_j B_j, A_j B'_j, A'_j B_j$, and $A'_j B'_j$ appropriate to a single pair of photons (pair "j") under the four angled polariser conditions. Following notation common in physics, we denote by $<AB>$ the average value of the column elements headed AB. That is $<AB> = (1/N) \sum_{j=1}^{N} A_j B_j$, and similarly for $<AB'>$, $<A'B>$, and $<A'B'>$. (The angled bracket notation is sometimes confused to denote an expectation, an issue we shall discuss in Chapter 6.)

Our first task will be to create a matrix representing a sequence of all four polarisation products on individual photon pairs, in keeping with the prescriptions of quantum theory. This can be done in a number of ways and will necessarily be tentative, because as mentioned, quantum theory does not prescribe a unique complete joint distribution over the values of A, A', B, and B' for a single pair of photons. This feature has been recognised since the insistent commentary of Fine (1982). Rather, the theory merely identifies the joint distributions for any *pair* of quantities that might be observed in an actually chosen experimental design, either $(A, B), (A, B')$, (A', B), or (A', B'). Any one of these pairs could be observed in an actual experiment, and quantum theory specifies probabilities for the possible paired outcomes of each. However, neither can the four experiments be conducted simultaneously, nor does the quantum operator specifying observation probabilities for any pair of them commute with the operators identifying probabilities for the other pairs. This precludes their joint assessment theoretically. Nonetheless, quantum theory does imply precise convex bounds on a joint distribution for all four quantities, deriving from its coherency with the four distributions on pairs that the theory does provide. These are the bounds we generated from a computational application of de Finetti's fundamental theorem of probability in Chapter 1.

To provide for a simulation that at least coheres with quantum theoretic prescriptions, we are required to conduct its operation in four stages. Each stage begins by generating a pair of polarisation quantities, either $(A, B), (A, B'), (A', B)$, or (A', B'), using recognised quantum probabilities. Whichever two are generated, the simulation routine will continue to generate the other two of the four (A, B, A', B') values for this photon pair, using conditional probabilities that will need to be discussed. I shall begin by randomly generating one million paired values for A and B on the basis of the recognised quantum probabilities pertinent to the observable paired polarisation angle $(\mathbf{a}, \mathbf{b}) = -\pi/8$. An ensuing generation of A' and B' will then be described in complete detail using five different alternative methods. The same strategy will then be followed again in a second stage, but beginning with the generation of (A, B') and continuing to generate the corresponding values of A' and B in the same way ... and so on, beginning subsequently with (A', B) and then with (A', B'). Once all four stages are completed, each to generate five full arrays of one million rows of prospective gedankenvectors, we will find what the sampling procedure touted by Gill and the others will yield. Let's go!

Generation of the pair (A, B) would begin using probabilities

$$P(A = +1 \ \ and \ B = +1) = \tfrac{1}{2}cos^2(-\pi/8) = .4268$$
$$= P[A = -1 \ and \ B = -1]$$

along with

$$P(A = +1 \ and \ B = -1) = \tfrac{1}{2}sin^2(-\pi/8) = .0732$$
$$= P[A = -1 \ and \ B = +1] .$$

Such a determination of the A and B values for a specific row of polarisation observations could be determined by experimentation as well, and this has indeed been performed many times. It has been decisively demonstrated that experimental polarisation observation frequencies do mimic those prescribed by the probabilities of quantum theory. There is no dispute about that. However, a determination of corresponding values of A' and B' for this same photon pair is required in order to implement Gill's sampling scheme. This can only be achieved by simulation, orchestrated by means of conditional probabilities that at least would cohere with specifications of quantum theory. These cannot not be definitive, because the theory does not really specify a complete unique distribution over all four observable quantities. They are only observable in pairs of two of them. Nonetheless, values for B' and A' can be generated as companions for the values of A and B in several ways, using conditional probabilities that *are* designated by quantum theory as a base. However, these will need to be chosen with care, which we shall clarify.

Conditional quantum probabilities do provide that

$$P[B' = +1|A = +1] = cos^2(-3\pi/8) = .1464,$$

$$\text{and} \quad P[B' = +1|A = -1] = sin^2(-3\pi/8) = .8536,$$

along with

$$P[A' = +1|B = +1] = cos^2(\pi/8) = .8536,$$

$$\text{and} \quad P[A' = +1|B = -1] = sin^2(\pi/8) = .1464.$$

Although using such a probabilities to simulate the generation of values for A' and B' in tandem with those simulated initially for A and B *would surely cohere* with the active prescriptions of quantum theory, they *would not be implied* by these limited prescriptions. This proviso requires a brief remark which will not detain us long, but it will need to be addressed again in discussion following the display of our simulation results.

All readers will be aware that any tendered joint probability mass function for any four quantities, including such as those considered here, can be factored into the form

$$f(A, B, A', B') = f(A, B) f(A'|A, B) f(B'|A, B, A'). \quad (1)$$

Replacing the second factoring conditional mass function $f(A'|A, B)$ with $f(A'|B)$, and the third factoring function $f(B'|A, B, A')$ with $f(B'|A)$ would both be permissible as a coherent factorisation. This would embed a feature of conditional independence into an imagined joint distribution, so as to yield

$$f(A, B, A', B') = f(A, B) f(A'|B) f(B'|A). \quad (2)$$

However, such a reduction would *not* be a requirement. The latter two factoring mass functions in (2) *could well be different* from these. An alternative that we shall actively investigate, for example, would be to replace the third factor in (1), $f(B'|A, B, A')$, by a random combination of $f(B'|A)$ and $f(B'|A')$ after A' has been generated from $f(A'|B)$ in the second factor of (2). But we are getting ahead of ourselves, because there are even more convenient alternatives that will be investigated as well. Even these will not exhaust all possibilities.

147

The simplification of replacing $f(A'|A, B)$ in (1) with merely the conditional $f(A'|B)$ as the second factor of the joint mass function $f(A, B, A', B')$ when conditioning on *both* A and B *could* be appropriate to quantum behaviour at the designated polarisation angle $(\mathbf{a'}, \mathbf{b})$. However, this proposal ignores the conditioning result A which has been construed as already observed at the angle pairing $(\mathbf{a}, \mathbf{b})$ at the beginning of this simulation. The replacement conditional mass function attends only to the result B. While such a simplification does not appear unreasonable, who is to say? Recognised entangled behaviour of A' in a real experiment with that yielding B has already exposed quantum behaviour that is surprising to many. Who is to say that the behaviour of A' would not be somehow entangled with its simultaneous prospect at A as well? Inaccessible to observation, quantum theory neither confirms nor denies any such suspicions. Similar issues would cloud the blithe replacement of $f(B'|A, B, A')$ in the factorisation with $f(B'|A)$ in the first simulation strategy we shall investigate. The important feature to recognise in the report of the simulation we shall now continue, is that the details we engage would cohere with the prognostications of quantum theory, but they are not required by them.

Deferring further discussion until we have produced the *complete* results of a simulation we are now only beginning, let's examine the results that appear so far. We have outlined heretofore only the first part of a four-part simulation project which begins with the quantum generation of (A, B). We shall continue subsequently to construct simulations that begin with generation of the pairings (A', B), (A, B'), and (A', B') once we get our bearings. Because we shall eventually compare and use the results of all four, they are now printed together in succession, as the four-part Table 1. But we shall discuss results of only the first one before outlining procedures for constructing the others.

Results of the first stage of simulation

The first Table 1(A, B) displays average results from 10,000 iterations of five distinct simulation runs on one million pairs of photons ejected toward stations A and B as imagined in the Aspect-Bell gedanken setup. The rows of the Table present results generated using five different factorisation procedures. The simulation generators designated as **AB1** and **AB2** *both* involve the simplification

148

Table 1(A, B). Average polarisation Product Simulations Initiated with (A, B)

PolsnProduct→ SimGenerator↓	< AB >	< AB′ >	< A′B >	< A′B′ >	< s >
AB1	.7071	−.7071	.7071	−.2071	1.9142
AB2	.7071	−.7071	.7071	−.3536	1.7678
A*B	.7071	−.1768	.7071	.1768	1.7678
AB*	.7071	−.7071	.1768	.1768	1.7678
A*B*	.7071	−.4420	.4419	.1768	1.7678

Table 1(A, B'). Average polarisation Product Simulations Initiated with (A, B')

PolsnProduct→ SimGenerator↓	< AB >	< AB′ >	< A′B >	< A′B′ >	< s >
AB′1	.7071	−.7071	−.2071	.7071	1.9142
AB′2	.7071	−.7071	−.3536	.7071	1.7678
A*B′	.1768	−.7071	.1768	.7071	1.7678
AB′*	.7071	−.7071	.1768	.1768	1.7678
A*B′*	.4420	−.7071	.1768	.4419	1.7678

Table 1(A', B). Average polarisation Product Simulations Initiated with (A', B)

PolsnProduct→ SimGenerator↓	< AB >	< AB′ >	< A′B >	< A′B′ >	< s >
A′B1	.7071	.7929	.7071	.7071	1.3284
A′B2	.7071	.3536	.7071	.7071	1.7678
A′*B	.7071	−.1768	.7071	.1768	1.7678
A′B*	.1768	−.1768	.7071	.7071	1.7678
A′*B*	.4420	−.1768	.7071	.4419	1.7678

Table 1(A', B'). Average polarisation Product Simulations Initiated with (A', B')

PolsnProduct→ SimGenerator↓	< AB >	< AB′ >	< A′B >	< A′B′ >	< s >
A′B′1	−.2071	−.7071	.7071	.7071	1.9142
A′B′2	−.3536	−.7071	.7071	.7071	1.7678
A′*B′	.1768	−.7071	.1768	.7071	1.7678
A′B′*	.1768	−.1768	.7071	.7071	1.7678
A′*B′*	.1768	−.4420	.4419	.7071	1.7678

of the imagined mass function $f(A, B, A', B')$ as factored in the required equation (1) to that prescribed in (2). The other three involve different agreeable modifications. All procedures begin with the generation of values for A and B using recognised quantum probabilities, applying seeds that are generated from a uniform distribution over $[0, 1]$. Then each procedure generates companion values for A' and B' from a different simplification of the factorisations we have mentioned. We shall describe and label them now.

The procedure **AB1** generates B' from its conditional distribution given the value of A so generated, according to (2), using a second uniform random number. This is followed by the generation of A' from its conditional distribution given B again using *the same* second number as its seed. Alternatively, procedure **AB2** uses different uniform random seeds in determining its B' and A'.

The upshot of this distinction for the simulations is that **Sim-Generator AB1** induces a correlation into its generation of B' with A' by using the same random seed for their conditional distributions given A and given B respectively. The generator **AB2** does not, using different random seeds to generate each. Either would be permissible by the abstemious prognostications of quantum theory. Quantum theory specifies a joint distribution precisely for the results of an experiment generating the polarisation pair (A, B), but it is resolutely silent regarding a joint conditional distribution for (A', B') given (A, B). We have some liberty in simulating joint values for (A, B, A', B') while honouring the prescriptions of quantum theory, and we have taken this liberty in distinguishing the generation processes **AB1** and **AB2**.

We then pursue this liberty further in alternative generating schemes labelled $\mathbf{AB^*}$, $\mathbf{A^*B}$ and $\mathbf{A^*B^*}$, to be discussed shortly. However, we should remark firstly upon the fact that *all five* of the simulation routines reported in Table 1(A,B) yield an average value of <s> within the bounds of $[-2, +2]$ required by Bell's inequality. Notably, they are also all within the bounds of $(1.1213, 2]$ identified in Chapter 1 as the bounds specified by quantum theory.

A further remark should be added as well. The numbers displayed in the Table derive from averages over 10,000 iterations of the polarisation product columns of the *entire* $N \times 4$ matrices computed for the simulations. If one were to select randomly a single column element from each of the 10^6 rows of the simulated matrices, and compute the sampled column averages from these, the results would differ from those appearing in the Table by only one unit at the fourth decimal place, just as Gill had surmised.

A final remark pertinent to the schemes **AB1** and **AB2** is that both of them produce average polarisation product values $<AB>$, $<A'B>$, and $<A, B'>$ with magnitudes on the order of .7071, which equals the rounded value of $1/\sqrt{2}$. It is only the magnitude of $<A'B'>$ that is different according to the two generators. Enough said for now, as this peculiarity will become clarified only by the end of the complete four-stage simulation.

A concluding explanation concerns the rows of Table 1(A,B) labelled $\mathbf{A^*B}$, $\mathbf{AB^*}$, and $\mathbf{A^*B^*}$.

The extension of the simulated observations of (A, B) to include A' using $f(A'|A, B)$ as equal to $f(A'|B)$ at the second step of simulation then begs the question of whether to generate the final value for B' at the third step using the quantum conditional mass function of $B'|A$ as we have in procedures **AB1** and **AB2**, or from that of $B'|A'$ using the value of A' that has now also been generated at step two. A similar quandary would arise if we were first to simulate a value for B' before the determination of A'. This might then be produced from a distribution conditional on either B or on B'. These strategic choices are engaged variously in the simulation results shown in the final three rows of Table 1 which include the terms $\mathbf{A^*}$ and $\mathbf{B^*}$ in their names, as explained now.

All simulations reported in Table 1(A,B) begin with a selection of values of A and B. Proceeding further, the **SimGenerator** labelled $\mathbf{A^*B}$ continues by immediately generating a value for A' from the quantum conditional distribution for $A'|B$. It then selects a value for B' from a random 50/50 choice of a selection from the conditional distributions of $B'|A$ and of $B'|A'$ using the value of A' so generated. As an alternative, the simulation labelled $\mathbf{AB^*}$ merely reverses the order of this process. It first determines the value of B' from its conditional given A, and then selects a generation of A' from a mixture of its two quantum conditionals, firstly that given the startup value of B, and then given this conditionally generated value of B'. Finally, the simulation labelled $\mathbf{A^*B^*}$ derives from a random 50/50 choice of selection for the pair (A', B') from the two alternative procedures, $\mathbf{A^*B}$ and $\mathbf{AB^*}$.

The five rows of results displayed in Table 1(A,B) deserve another comment before discussing the remaining three Tables. In the rows deriving from generators $\mathbf{A^*B}$ and $\mathbf{AB^*}$, only *two* of the average polarisation products resolve to values on the order of .7071 (which is the rounded value of $1/\sqrt{2}$), while in the simulation $\mathbf{A^*B^*}$ only the single average $<AB>$ resolves to this value. Just saying. There is further intrigue that might be noticed in the comparative results of the simulations $\mathbf{AB2}$, $\mathbf{A^*B}$ and $\mathbf{AB^*}$, but a discussion in depth would distract from larger issues that will become clearer by moving directly on to the next three dimensions of our simulation.

Recognising that there is nothing sacrosanct about the polarisation angle $(\mathbf{a}, \mathbf{b})$ to begin the quantum simulation with the generation of a pair (A, B) as we have in producing Table 1(A,B), we are impelled now to generate a complete array of polarisation products that needs to weather examination in a quantum assessment of Bell's Inequality.

Completing the simulation assessing Bell

To complete the generation of a population of simulated polarisation products for photon pairs at all four angled designs, the structures of five **SimGenerator** procedures described in Section 3 were repeated three more times, each with one basic variation. The results in Table $1(A, B)$ derive from a procedure beginning with the production of quantum observation pair (A, B) and with subsequent generation of the concomitant A' and B' in five different ways. The remaining three Tables derived from following the same procedures structurally, but were *initiated* with a quantum theoretic generation of a *different* pair of polarisation observations, either $(A', B), (A, B')$ or (A', B'). Whichever pair initiates the run, the remaining two simulated observations were then generated according to the same five process types we have used in producing Table 1(A,B).

The results of each procedure appear in a Table of its own. We have designated them as Tables $1(A', B)$, $1(A, B')$, or $1(A', B')$, each initiated with its entitled polarisation pair. The row titles within each Table are virtually identical, modified only to designate the initiating pair of polarisation observations as $AB', A'B$, and $A'B'$. I suggest that you now examine the Tables for yourself, before I comment. You will notice structural similarities that arise in the originating Tables $1(A, B)$ and $1(A', B')$ and those that arise in Tables $1(A', B)$ and $1(A, B')$. Peculiarities of the results distinguishing these pairings derive from the angle size of $-3\pi/8$

between the polariser directions $(\mathbf{a}, \mathbf{b}')$ and the size of $\pi/8$ or $-\pi/8$ that characterise the others. You may review these angles in the depiction of Figure 2 in Chapter 1, page 5.

One feature of the four Tables of results is crucial to issues concerning the supposed empirical validation of the defiance of Bell's inequality. Each of the four component Tables 1(A^*, B^*) displays an average value $<\mathbf{A}^*\mathbf{B}^*>$ *for the polarisation product column A^*B^* that generates it* as equal to its corresponding quantum theoretic expectation:

$E(AB) = E(A'B) = E(A'B') = 1/\sqrt{2}$, while $E(AB') = -1/\sqrt{2}$.

These values appear in the four Tables of averages with magnitude .7071 equal to the expectation of the polarisation product that initiates the simulation. The averages of the other columns in each Table do not always do so.

What is substantive about this feature of the simulation results? The claim of the community of physicists who defer to arguments presented by Aspect/Redhead/Mermin/Gill is this: the random selection of the polariser direction pair $(\mathbf{a}^*, \mathbf{b}^*)$ in Aspect's CHSH experimental design suffices to constitute the resulting observed polarisation product (A^*B^*) as a random multinomial choice of an element from a spreadsheet row of polarisation products from a gedankenexperiment. After all, the direction is determined between the moment the photons leave their source atom and the moment they arrive at their polarisers. According to the procedure explained by Gill, the spreadsheet is supposed to derive from a matrix of polarisation observations on a single photon pair at all four polarisation angles, just as we have constructed. However, experimental results on photon pairs at single polariser angle pairs are *not* selected randomly from among these. Well, why not?

True, multinomial random choice does characterise selection of a polarisation angle that would generate each of the four Tables we have examined. Having selected a polariser angle pairing, however, a real experiment is limited to produce an observed polarisation product only from the column corresponding to the angle that generates the Table. There is no chance at all that a real observation is selected randomly from a row of a gedanken matrix such as we have produced. Computing an estimated "$\hat{E}(s)$" matrix from averages of Aspect-type observations misconstrues the empirical evidence relevant to quantum theoretic prescriptions pertinent to Bell.

When we observe the results of actual experimentation randomised over the four angle pairings, collected in four appropriate columns, there is no restriction that the average value of $<s>$ so determined as $<AB> - <AB'> + <A'B> + <A'B'>$ need lie within the interval $[-2, +2]$. As is widely known by now, in case after case of investigation it does not!

The source of the error in the touted empirical defiance of Bell's inequality is that the estimation procedure followed ignores the symmetric functional relations inhering among the row elements of the gedanken matrix we have simulated. These also underlie the well-recognised restriction of the CHSH form of Aspect/Bell's quantity "s" to the values of only -2 or $+2$. Moreover, they were central to the linear programming procedures we followed in Chapter 1 to determine the bounds on $E(s)$ as the interval $(1.1213, 2]$, actually prescribed by quantum theory.

To investigate the results of sampling procedures designed to evaluate the prescriptions of quantum theory pertinent to Bell, one would need to sample elements from the full spreadsheet of polarisation products we have constructed. These use all four pairs of polarisation observations to start the simulation. This would allow us to instantiate the procedure specified by Gill (2014). Merely to use observations generated for a single relative polarising angle does not suffice.

Average simulations assessing Bell

To consider *a population of rows* of polarisation products that constitute the type of matrix that Gill considers to be his spreadsheet, we need to generate *complete rows* of gedanken observations *based on all four different pairs of starting generators*. Doing this would amount to constructing a $4N \times 4$ matrix, composed of a stack of the four $N \times 4$ matrices we used to produce the results we have already displayed. Averaging the components of *this* stacked matrix would amount to averaging the results in the four matrices already displayed as Tables 1(A*,B*). Such an average is presented as Table 2 below. Because the size of the component matrices is so large (one million rows each), uniform multinomial sampling of one element from each row of this large matrix would yield sampled column averages that hardly differ from the column averages appearing herein. The results are perhaps surprising.

Table 2. Average polarisation Product Simulations over ALL ANGLE INITIATIONS

PolsnProduct→ SimGenerator↓	<AB>	<AB'>	<A'B>	<A'B'>	<s>
One Rand	.4786	−.3321	.4786	.4785	1.7678
Two Rands	.4419	−.4419	.4419	.4419	1.7678
A,A' for B*	.4419	−.4420	.4419	.4419	1.7678
B,B' for A*	.4419	−.4419	.4419	.4419	1.7678
Mixed A*,B*	.4419	−.4419	.4419	.4419	1.7678

The rows of the individual Tables numbered $1(A^*, B^*)$ were labelled with names identifying their starting pair of generators. The five rows of Table 2 are labelled with names indicating only the specific procedure used to complete the four polarisation products. Each row of Table 2 is the average of the corresponding rows in the four component tables of Tables $1(A^*, B^*)$, initiated with a different one of the various polarisation pairs (A^*, B^*). The *first row*, for example, reports the averages $<AB>$, $<AB'>$, $<A'B>$, and $<A'B'>$, *but averaged over the first rows of all four* initiating polarisation product pairs in Tables labelled $1(A^*, B^*)$. Since these first rows are all generated following a procedure that involves only one random number for each of the two polarisation observations remaining after the simulation of (A^*, B^*), the row of Table 2 that corresponds to this average is labelled **One Rand**. Appropriately, the second row of Table 2 is then labelled **Two Rands**. The remaining row labels $(\mathbf{A,A'}$ **for** $\mathbf{B}^*)$, $(\mathbf{B,B'}$ **for** $\mathbf{A}^*)$, and (**Mixed** $\mathbf{A}^*,\mathbf{B}^*)$ are merely suggestive of the more ornate **SimGenerator** procedures used to produce the final three rows in each of the four component tables of **Table 1**. These rows were each averaged to produce Table 2.

Each row of Table 2 is the average of the corresponding row of the four Tables $1(A^*, B^*)$. As such, it is the average of ten thousand replications of procedures yielding the column averages of a matrix sized four million by 4. If averages were computed alternatively from independent multinomial samples $M[1, \mathbf{p}_4 = (.25, .25, .25, .25)]$ selected from the four million rows, the results would differ from the results shown in Table 2 at most by one unit in the fourth decimal place, just as Gill had suspected. However, the results are quite different from what he suggested. Your examination may surprise you. Let's discuss them.

Concluding commentary

To begin, it may appear as a surprise that *none* of the average polarisation observations appearing in the columns of Table 2 is near to the value of $1\sqrt{2} = .7071$. Do not be surprised! The widely reported averages of experimental observations at individual polarisation angles that do regularly arise near to .7071 have nothing to do with the context of Bell's inequality. This pertains to a gedankenexperiment that cannot be implemented, requiring simultaneous observation of polarisation products on a pair of photons at four differing relative polariser angles. What does appear intriguing is that the average value $<s>$ displayed in the final column of Table 2 equals 1.7678 under all five of the **SimGenerator** procedures we have orchestrated. Oddly, this is also the value that arose in our initial expository simulation analysis, designed in still another way as reported in Chapter 1. Mimicking the actual empirical procedure followed by Aspect, this presumed independent generation of any three polarisation product components of Bell's quantity according to quantum probabilities. However, it then completed the fourth merely by applying the restriction that the value of s can equal only either -2 or $+2$. This numerical result turned out to equal a rounded value of $(2.5)/\sqrt{2} = 1.7678$.

It is intriguing that this same number also arises in the results of every simulation procedure reported in Table 2. Nonetheless, this result cannot be considered to be *definitive* as a quantum theoretic expectation of s. For a full interval of possible values for $E(s)$ within $(1.1213, 2]$ cohere with recognised quantum prescriptions. Interestingly, several other specific values in this interval can be identified for some notable distributions among the many that would cohere with quantum theory. I can merely report here that further Monte Carlo computations identify the *maximum entropy* sourced expectation of Bell's quantity that would cohere with the four product expectations provided by quantum analysis. Surprisingly, the order of magnitude of this expectation is quite different still, a value equalling 1.1296. This result have recently been published in the *Journal of Modern Physics* (Lad, 2023).

The bottom line to end this chapter is that claims to empirical evidence of the defiance of Bell's inequality garnered from average observations of polarisation products at four distinct relative angle settings are simply mistaken.

Before we conclude this book in Chapter 7 with a re-evaluation of John Bell's (1965) original article challenging the argument of EPR, we shall first take liberty to discuss some extraneous aspects of physicists' attitudes toward quantum probabilities that have coloured their mistaken understanding. This is the project of Chapter 6.

Chapter 6

ON PROBABILITY
AND QUANTUM PHYSICS

Having attended exclusively to technical matters in the initial five chapters of this book, this chapter adjusts our focus to some interpretative matters of quantum physics. They do all border on technical issues. After initial reflections on twentieth-century fascinations of quantum theory with matters of probability, I present a terse review of the formal structure of mathematical expectation as it has been developed following leading work of Bruno de Finetti. He recognised it as a unifying concept that includes probability, a formalisation of uncertain personal judgements, and designated it as "prevision". My display is followed by an exhortation to recognise fundamental differences between expectations and averages. Further dissection of distinctions between joint, conditional, and marginal probabilities will be found relevant to common interpretations of quantum theory that I challenge as unwarranted and misguided. These will be clarified by addressing the feature of exchangeability of probability assertions, a property of symmetric assertions regarding vectors of observable quantities. It is commonly appropriate to applications of scientific concern at every scale of experience, classical or quantum. We will find that this is the feature of scientific attitudes toward particle behaviour that physicists have objectified in their fabled discovery of "quantum entanglement". To conclude this chapter, an exposition of the surprising universal inherence of complex amplitude structures within *all* probability distributions puts paid to claims of the uniqueness of quantum probabilities in deriving from such underlying specifications.

The preceding chapters have exhibited details of a technical error in the commonly accepted assessment of mathematical expectation for a speculative physical quantity, an error which has led quantum theorists to some astounding and bizarre claims about the random behaviour of matter at minute scales of activity. While the theory of probability itself has been a controversial subject over the entire course of its historical development, stemming largely from differences in attitudes of practitioners toward the meaning of its premises, the error we have addressed does not rely on any specific position concerning relevant issues. The error rests on purely technical matters, no matter how one might interpret the meaning of the propositions involved. It arises from the neglect of four symmetric functional restrictions inhering in the summands of a linear combination of four polarisation products in a hypothetical experiment. Crucial to understanding the error is recognition that while the summands of the linear combination are each observable, their linear combination is *not* observable. The experiment that would identify the quantity cannot actually be conducted. Quantum mechanics does not specify anything precisely about unobservable phenomena. Nonetheless, it can specify some bounds on what could make sense based on its precise specifications regarding related actual physical phenomena.

To be sure, the problem addressed by the gedankenexperiment underlying the CHSH form of Bell's inequality has developed from, and indeed revolves about, differences in theoretical attitudes toward the meaning of probability. Einstein's view was that the probabilities prescribed by quantum theory represent scientific uncertainty regarding observable quantum phenomena in tandem with unspecified supplementary variables. This was confronted by the more popular view among physicists that the probabilities represent properties of physical systems themselves, typically conceptualised as ontic physical "propensities" in the structure of matter itself.

The mathematical theory of probability and statistics is now on the rebound from some seventy-five years of intense scrutiny of its complete foundations. While matters of quantum physics have been recognised as relevant to the discussions, mathematical interest in the axiomatisation of the theory and its implications for empirical practice has been pursued in forums somewhat removed from those engaged by physical theorists. To be sure, Bruno de Finetti did devote several appendices in the second volume of his *Theory of*

Probability to matters of the quantum formulation, and it is clear that his position on these matters developed throughout his intellectual life. See the exposition of Zabell (2009). In the same era, Schrödinger (1947) wrote two papers in which he developed ideas quite in keeping with de Finetti's constructions, though evidently independently of them, proposing that probability can represent a graded assertion of uncertain knowledge. However, much more influential in circles of physics were the works of Hans Reichenbach (1944, 1949) who had written his *Philosophic Foundations of Quantum Mechanics* in concert with his own *Theory of Probability*, and the extremely influential writings throughout 1930-1950 of Richard von Mises. These characterised probability as the study of aggregate phenomena and recurring events of stable frequency. There has developed by now an extensive literature discussing interpretations of quantum probabilities, well reviewed in the illuminating volume of Jaeger (2009) on *Entanglement, Information, and the Interpretation of Quantum Mechanics*. Nonetheless, the most general forum for the discussion of foundational issues in probability *per se* has been literature deriving from the energies of L.J. Savage, whose 1954 *Foundations of Statistics* was of monumental influence.

Despite the active spirited participation of Edwin Jaynes at the NBER-NSF meetings at which the developments were regularly engaged during the 1970s and 80s, to some extent issues relevant to the probabilities of quantum theory have been aloof from these foundational discussions, dominated as physics has been by objectivistic attitudes among most practitioners. These rely on the laws of large numbers to support the notion that the probability of an event is the average number of its random occurrences in repeated instances of the opportunity. It is within this somewhat narrower context that probabilistic quantum controversies and interpretations have been addressed. In contrast, foundational literature on statistical probability has centred more typically on the characterisation of procedures for applied statistical inference. In this vein, the statistical assessment of experimental research in physics has remained fairly naive. Indeed, Aspect's estimation procedure for the expectation of the CHSH quantity "s" relied on the very old-fashioned and naive "method of moments". More recent claims to the defiance of Bell's inequality by subsequent experimental activity are commonly reported using the completely discredited reliance on p-values. No worries, as this is common statistical practice.

Happily, there have been notable and refreshing inroads of subjective Bayesian ideas into the discussion of physics itself, led by writings of Christopher Fuchs and colleagues. While there is by now an active coterie of quantum theorists who do work in this vein, the larger community of physicists has remained sceptical and unmoved. A central reason for this lies in the domain of the mistaken understanding of Bell's inequality that I have untangled in the first five chapters of this book.

In this penultimate chapter, I highlight specific crucial features of probability that are misconstrued in the currently orthodox discussions of quantum physics. I will focus on the resolution provided by constructive subjectivistic mathematical analysis for some quandaries of quantum theory and its supposed support for mysterious behaviour of matter at its tiniest scale. I need to introduce this programme with some historical and philosophical remarks. I fancy that my analytical reassessment of the touted defiance of Bell's inequality permits me to make them. You may find the remarks upsetting or challenging, but they will not take long, soon to be followed again by purely technical remarks.

Arguably the most profound insurgence of subjectivistic understanding of all matters statistical has arisen from the work of Bruno de Finetti (1906-1986). His contributions began flourishing exactly in the years of the 1930s when Albert Einstein (1879-1955) was actively challenging the ontic notions of quantum probabilities promoted by the Copenhagen varieties of theorists. Einstein's opinion that the probabilistic formulations of quantum theory derive from our uncertainties regarding unknown features of physical activity at the atomic scale was not well regarded by other developers of the theory at the time. Their objectivistic point of view has largely won the day in the physics community through subsequent decades. Notable in its support has been the apparent defiance of Bell's inequality by quantum probabilities, which I have debunked in my demonstrations herein. On this basis, Einstein's views have been relegated to the sideline as those of a peculiar dissenter.

The concerns of de Finetti were largely mathematical. His point of view as a mathematician was essentially constructive, proposing that the subject of mathematics is not merely an arbitrary formalism. Rather, it constitutes a language for talking about assertions that actually mean something operationally. And what is the subject matter of probabilistic language? It is the coherent implications of uncertain assertions that anyone may make regarding the observable values of operationally defined quantities. Its programme is to provide a formalism whereby the information contained in new experimental observations can be codified to support inference from them regarding further observations yet to be made.

On the sideline relative to the burgeoning mainstream developments of statistical hypothesis testing and procedures of probability "estimation" through the mid and late twentieth century, de Finetti's work has developed among sympathetic researchers to provide the basis for a completely different tradition of statistical activity that is known as "scored statistical forecasting". Though there may well be observable quantities we might wish to estimate, there are no probabilities to be estimated. Probabilities, rather, are numerical gauges of assertions representing uncertain knowledge about quantities we can observe.

De Finetti was well aware of Einstein's ideas, and of the controversies developing in the formulation of quantum theory. He wrote a dense commentary on such matters in the Appendices to his swansong text, *Teoria della probabilità*, translated into English in 1973/74. Yet, while his writings were substantive on formulaic matters, he did not comment on the propositions of Bell's inequality, nor does he appear even to have been aware of them. These were first proposed to the research community only in the mid-1960s. Recognition of their importance took some time to take on, for reasons well described in the fascinating commentary of Becker (2018). Bell, it should be known, had begun his own research programme feeling attracted to Einstein's ideas, and had hoped to provide their development with renewed vigour. He was then evidently more than surprised by the apparent implications of the inequality he had identified. Resigned to trusting his mathematics, he nonetheless surmised that the error embedded in its formulation would come to be discovered in due time. This was an opinion that he reaffirmed in a compendium of his writings in the late 1980s. Neither he nor Einstein seemed to have been aware of de Finetti's work, as I find no reference to it in either of their writings.

Nonetheless, the challenge presented in the EPR article, embellished with Einstein's allusion in a letter to a friend that he could not really believe that "the old one rolls dice", lies plum flush with formal developments deriving from de Finetti's work. These are largely ignored within the physics community, by practitioners who exhibit no qualms with allusions to stable frequencies of occurrence of random events as evidence for the existence of actual probabilities that generate them, unobservable in themselves.

As for me, fairly well conversant with the developments of mathematical probability theory and with argumentation surrounding them, I am an avid proponent of the operational subjective theory of probability championed by the insights and committed research career of Bruno de Finetti. Pondering the fantastic images of the galaxies produced by the Hubble/Webb telescopes and their interpreters, and even pondering the more mundane and concerning sociopolitical and ecological events of our days, I find it preposterous that scientists can speak glibly in terms of "repetitions of identical conditions" underlying their conceptions of objective probability as the generator of our physical experience.

The fact is, as commonly understood by the scientific consensus of our day, we are in the midst of an incredible explosion which has been propagating itself for some billions of years! Rather than standing somehow apart from it and somehow "measuring *it*", we *are* "it", at least a part of it. We seem to have emerged within it in time, and we are quite confident that we will disappear from it in time, surely as our own solar system collapses, if not before. This is neither a joke nor an insight that can or should be taken lightly. The so-called "measurement problem" of quantum physics, which has absorbed so much energy and concern over the past near century, derives from a world view that can no longer stand the scrutiny of serious scientists. That there exists some world different from us who stand apart from it, which we can manipulate but only measure with error, existing in some "state" whose measurements can be repeated, is a notion that derives from a now quite remote era of the mechanics of the industrial revolution. I do not propose any resolution of "the measurement problem". Rather, I propose its dismissal.

A different understanding of the project of physical science derives from addressing the simple proposal that something is happening. "What's happ'nin?", a perhaps now dated greeting of the streets, is the ultimate scientific question. We are not sure what it is, but we are quite evidently part of it. In the course of our living within it, we record summary observations of what the happening includes, and we call them "statistics". The statistics do not exhaust the many dimensions nor the possible descriptions of the happening, and they cannot possibly do so. That is what makes them mere statistics. Our participation in the construction of these recorded statistics is part of the statistical event itself.

Consider something as banal as the conception of "my height". Twentieth-century statistical characterisation of anatomical measurement presumes that I do have a "true height", typically denoted by a Greek letter, such as μ. Unobservable in itself, it is presumably measurable only with error, as described in notation by the equation $X_t = \mu + \epsilon_t$. Here the addend ϵ_t is proposed to represent an unobservable random error of measurement, whose stochastic properties are particular to the characteristics of the measuring device and the skill of the specific person who wields it. Different measuring devices and different administrators of the measurement are proposed to be characterised by different stochastic properties.

In contrast to such an understanding, I propose that I simply do not "have a height". I live a life. My body has been growing and shrinking all the time! If you would like to measure me with a yardstick, or whatever device, shortly after a hard day of my reconstructing the foundations of the house next door, this measurement would be a live summary recording of history, an unrepeatable activity. Knock yourself out! The measurement is what it is. It would be quite distinct an episode from your measuring me after a pleasant session of yoga, again even the same evening! It is preposterous to regard these as distinct repeatable measurements of a true unobservable height, sullied by random disturbances and measurement error. These musings are strikingly different from those which presume that every physical quantity "has a value" at all times, though it can only be measured with random error.

This is not the place to go into it all, because I'd like to go on to address what I regard as several specific statistical misconceptions involved in routine discussions of quantum theory. Suffice it merely to say for now, it is rather strange that the physics community has bought into such conceptions of unobservable quantities and their error-laden measurement. What is conceived as real by "realist" physics is the unobservable world of Greek letters. This conception relegates the fleshy Roman-letter world in which we live and participate, as the mere emanations of random variables which this ethereal world emits. So I will conclude this introduction by foreshadowing my contention: that the subjectivistic characterisation of the judgement to regard specific events "exchangeably", in its extended formulations, is the technically recognised substantive construct that allows us to understand the experimental observations of quantum physics. It formalises a conception of statistical inference from digitised observations that is appropriate to codify the achievements of all scientific activity in the empirical domain.

In the several sections of this chapter, I shall address statistical topics of expectations and averages, conditional probabilities and their meaning, the collapse of the wave function, the disturbance of a physical process induced by its measurement, repetitions of measurements, the property of exchangeable statistical assessments, and the derived position of probabilities as squares of their complex amplitudes. The discussions will serve to support the recognition that quantum probabilities are no more mysterious than any probabilities at all, serving merely to formalise our knowledge of observable physical processes about which the scientific community is uncertain.

Axiomatic probability theory, ... expectation or prevision

It is a sign of the times, indeed a sign of the entire era of twentieth-century mathematical formalism and its vestiges, that the incisive presentation of Jaeger (2009) on the the issues of entanglement, information, and the meaning of quantum theory, takes the famous axiomatic measure-theoretic scheme of A.N. Kolmogorov (1933) as the basis for its discussion of probability. The developments of probability in the nineteenth-century, riddled with confusion and perceived paradox, had transformed the early subjectivistic views of Laplace and Augustus De Morgan into the objectivistic reformulations of John Venn and beyond. This led the mathematical injunction for probability theory to take refuge by the next century in the set-theoretic formalism of Kolmogorov as a stabilising totem. Though he suggested as an interpretation the apparent conformity of probability with frequency, the clarification of semantic matters was never a prominent goal for him. Examination of the history is best rewarded in the excellent, appealing work of Hacking (1990) and the compilations of Gigerenzer et al (1989), ... and perhaps even my own take on the matter (Lad, 1996).

As prescribed by the proponents of formalist mathematics as prominent as Bertrand Russell and David Hilbert, the goal of the formalism is merely to verify a complete and consistent formulation of the syntax and content of a theory, meaning nothing in itself. Never mind the semantics. One might interpret the formalism however one might, and use it as seems practical and efficacious for arriving at useful results. Jaeger's compilation of the various views relevant to quantum theory seldom crosses the boundaries of this understanding, recognised as it was by the main players in the debate. As a professionally recognised stance, this strategy cannot be faulted in codifying the concerns of physics through the century and beyond.

Unknown to (or ignored by) many participants was a quite different formulation of mathematical probability that derived rather from a constructive program for the foundations of the mathematics, yielding a function-theoretic characterisation of the syntax for probability as a derived notion. The axioms of its mathematical construction are not merely formal, designed simply to yield whatever they might as a useful result. Rather, they formalise a meaningful activity of human assertion. With an equally long pedigree of creative thought, stemming from the *Ars Conjectandi* of Jacob Bernoulli, the alternative formulation makes axiomatic the character of expectation rather than probability. In Kolmogorov's formalism, the notion of expectation is a derived notion, secondary to the specification of a probability distribution. It was the constructivist line of thought that captured the imagination of de Finetti in the 1920s. It encouraged him to develop a programme of research activity related to, but quite distinct from, the formal axiomatics of Kolmogorov. Indeed, taking the semantics of the formulation as fundamental, de Finetti laboured through the century in the shadows of leading developments in statistical theory and practice. These were championed by such notable theorists and practitioners as Neyman, Pearson, Fisher, Mahalanobis, and Gnedenko.

Ever keen to debate fundamental issues, de Finetti wrote one spectacular analysis of the situation entitled "On the axiomatisation of probability". Published in Italian, 1941, and in English translation, 1972, it is not widely discussed seriously. In his swansong text, the professor lamented that "a dialogue between the deaf is not a discussion". (*Theory of Probability*, Section 1.3.2.)

For myriad reasons that will not detain us here, de Finetti coined the word "prevision" to denote the mathematical construction that would play the role of expectation in standard formulations of probability theory. Formally, all probabilities are expectations, but one can assert an expectation (prevision) without asserting any probabilities. Without any parenthetic remarks or digressions to divert us, the following Section presents a *terse* formal statement of how the construction proceeds, in the universally practical case of actual statistical observations, finite and discrete. You may read it as a primer, or you may merely peruse it to get an idea of the extent of its ornate and exhaustive formality. Alternatively, you may skip it completely and move on to the next Section, if this mathematical detail would be a distraction to your interests.

Prevision: a subjectivistic construction

Every statistical recording specifying a *quantity X*, identifies one of a finite discrete number of possibilities to summarise the observation of what is happening in some domain. This exhaustive list of possibilities is termed *the realm of X*: $\mathcal{R}(X) = \{x_1, x_2, ..., x_K\}$. More completely, every *vector of quantities* $\mathbf{X}_N$ identifies one of a finite, discrete list of vectors of such recordings $\{\mathbf{x}_{N,1}, \mathbf{x}_{N,2}, ..., \mathbf{x}_{N,K}\}$. The matrix of these possible observation vectors is called its *realm matrix*: $\mathbf{R}(\mathbf{X}_N) \equiv \mathbf{R}_{N,K}$.

Typically, you are uncertain in your knowledge of the component values of $\mathbf{X}_N$, and your assertion of the state of your knowledge about them is represented by your *prevision*. Your prevision for $\mathbf{X}_N$ (a motivated name extending the concept of "expectation", which is more limited) is a vector of numbers $P(\mathbf{X}_N) = \mathbf{p}_N$ that you assert to express formally your uncertain knowledge of the values composing $\mathbf{X}_N$. In asserting this vector, you profess thereby your indifference to any transaction that will yield you a *net gain* of $NG[\mathbf{X}_N, P(\mathbf{X}_N)] \equiv \mathbf{s}_N^T[\mathbf{X}_N - P(\mathbf{X}_N)]$. The units of this gain are understood to be units of utility, or small amounts of cash. The components of $\mathbf{s}_N$ may be chosen to be of either sign, understood to be small enough only so that the absolute scale of the net gain is small, that is, $|NG[\mathbf{x}_N, P(\mathbf{X}_N)]| \leq \mathcal{S}$ for every possible actual observation value of $\mathbf{x}_N$. This suitably small number $\mathcal{S}$ is called the *scale of your maximum stake*. It is specified formally only so as to avoid difficulties that might arise in your assessments, due to the non-linearity of your utility valuation for increasing amounts of money. Positive valued components of the vector $\mathbf{s}_N$ equal to s_i

denote the transaction involving your purchase of the value of $s_i X_i$ for the price of $s_i p_i$, which you would pay. Negative components of $\mathbf{s}_N$ specify the transaction involving your sale of the value of $s_i X_i$ in return for your receiving the price $s_i p_i$.

Your prevision vector is said to be *coherent* just so long as there can be found no vector $\mathbf{s}_N$ that would ensure your net gain $NG[\mathbf{x}_N, P(\mathbf{X}_N)]$ is negative for every possible observation value of $\mathbf{x}_N$. Coherent prevision assertions do not permit you to be taken as a "sure loser" in these net transactions, no matter what the observation recording happens to be. This is a technical condition of your assertions, generalising the principle of non-contradiction in two-valued logic. Geometrically, the separating hyperplane theorem implies that your assertion $P(\mathbf{X}_N)$ is coherent if and only if it lies within the *convex hull* of the columns of the realm matrix $\mathbf{R}_{N,K}$. Algebraically, prevision assertions are coherent if and only if they guarantee prevision acts as a *linear operator* on a linear space of functions of $\mathbf{X}_N$. That is, $P(\mathbf{c}_N^T \mathbf{X}_N) = \mathbf{c}_N^T P(\mathbf{X}_N)$ for any coefficient vector $\mathbf{c}_N$.

Any quantity whose realm contains only the possibilities 0 and 1 is called an *event*. Events are not categorically different from quantities with more extensive realms. A prevision for an event may also be called a *probability*, which of course is also an expectation in the now standard Kolmogorov formulation. (The expectation for any event is equal to its probability.) The coherency of prevision ensures the satisfaction of Kolmogorov's axioms, applied to finitely additive probabilities. Furthermore, coherency ensures the validity of the representation of an expectation as a derived notion within the axiomatisation. Since we can represent any single quantity X by $X = \sum_{i=1}^{K} x_i (X = x_i)$, which constitutes a linear combination of events, its prevision equals $P(X) = \sum_{i=1}^{K} x_i P(X = x_i)$ according to the coherency condition of linearity. This formulation of coherent prevision unifies the characterisation of prevision (expectation) and probability. There is no need for two distinct structural formulations. All probabilities are previsions. Not all previsions are probabilities, simply because not all quantities are merely events.

A *conditional prevision* is a price valuation for a conditional transaction, asserted operationally. Asserting $P(X|E)$ involves the specification of a price value at which the quantity X is freely exchanged (as a purchase or a sale) for the price $P(X|E)$ with a side condition: if the event E eventuates to be observed equal to 1, then the transaction arrangement is engaged, whereas if E is observed to equal 0, then the tentative transaction is cancelled. In the former case, the resulting net gain may be positive or negative. In the latter case, the net gain is surely equal to 0.

If someone asserts a conditional prevision value $P(X|E)$ as a number, we can define a *conditional quantity* relative to this value as $(X|E) = XE + (1 - E)P(X|E)$. By this definition it is specified as a number whose numerical value formalises the result of the "contract" we have just specified. The linearity of coherent prevision then requires that $P(XE) = P(X|E)P(E)$ if the values of $P(XE)$ and $P(E)$ are to be asserted concomitantly, with no limitation on the value of $P(E)$ other than residing in the closed interval $[0, 1]$. Of course, someone may assert a value for $P(X|E)$ without asserting $P(XE)$ or $P(E)$. As a development distinct from this, the Kolmogorov formalisation of probability requires the *definition* of conditional probability for an event E_2 given E_1 via $P(E_2|E_1) \equiv P(E_1 \cap E_2)/P(E_1)$ as long as $P(E_1) \neq 0$.

The comments I shall now make regarding issues of probability that are misconstrued in many discussions of quantum theory, are simple, though perhaps long-winded. Having avoided the use of formal subjectivistic language and notation in this book until this point, I will now begin to use them unabashedly in the rest of this chapter. Its usage is by now natural to me. However, my comments pertain to mathematical probability, in whatever way it is formalised. When I write something like "if you would assert your prevision as $P(X)$ or $P(E)$ ...", if you are an objectivist, you may merely read this as saying "if you would like to entertain an expectation $E(X)$ or a probability $P(E)$, and/or estimate them as $\hat{E}(X)$ or $\hat{P}(E)$ ...". Let's go.

Expectations and averages

A probability is not an average. Full stop. Neither is an expectation. If one would like to consider a sequence of statistics, $X_1, X_2, ..., X_N$ of any length, then their numerical average is called a statistic too: $\bar{X} \equiv \sum_{i=1}^{N} X_i$. If you do not know what are the values of the X_i, then typically you will not know the value of $\bar{X}$ either. (Of course you may know the value of $\bar{X}$ without knowing the value of each X_i.) You may assert your previsions for the various possible values of X_i, and avow too the prevision $P(\bar{X})$ that coherency implies. Or you may assert your $P(\bar{X})$ and live with the restrictions on your cohering previsions for the components of $\mathbf{X}_N$ that are implied. Moreover, you may assert whatever probabilities you wish for the possible values of $\bar{X}$.

Neither a probability nor an expectation *is* an average. It is fair enough for a professor to write down the formula $E(X) = \sum_{i=1}^{N} x_i P(X = x_i)$ and to tell students, "See, an expectation is something like a weighted average of the possible values of X." There is no quibbling with what might help get an idea across as an introductory notion. However, there is no "average value" of X. There is only one value of X, and it will be observed to be what it is. There may well be other quantities with different values about which you may entertain similar uncertain ideas, other X's if you will, but that is another matter about which we shall comment shortly. What will not fly is to claim that a probability or an expectation *is* an average.

It is just incorrect for anyone to claim that the probabilistic prescriptions of quantum theory are statements about averages, rather than statements about individual events. Even if you are Paul Dirac! See the discussions in Jaeger (2009, pp 61 and 119).

The laws of large numbers are not propositions concerning facts of Nature. Rather, they are prescriptions that govern the coherent implications of your assertions about some quantities of which you are uncertain, for other quantities that are defined by functions of them. The weak law of large numbers, for example, in its simplest (and largely irrelevant) form, says merely that if you regard a sequence of N events independently with the same probability p for any finite sequence length, then your concomitant probabilities that the average $\bar{E}_N$ of these events differ from p by more than an exceedance, K, must become arbitrarily small as you consider longer sequences of larger sizes of N. (My parenthetic allusion to the "irrelevance" of the construal of independent events shall remain merely a provocative tease until we discuss the property of exchangeable assertions.) This theorem is an injunction concerning the coherency of your concomitant independent and identical probability assertions for events in a sequence, and for their averages.

When probabilities are construed as unobservable ontic propensities, the theorem is a proposition concerning limiting relationships between averages and a purported entity (the identical "p") that can never be observed. It is surely *not* a statement that provides a foundation for understanding the probability to "be" an average. Such construals were challenged even by the eminent Andrei Andreevich Markov himself in the early days of his work on limit theorems. Challenged that the law of large numbers substantiates the interpretation of probability as a law of mass phenomena (the average of many independent events), he merely pointed out that the probabilities of the individual events are part of the suppositions of the theorem. Enough said.

Conditional probability and observation

My remarks are material to three types of agreed quantum probabilities pertaining to the polarisation behaviour of the paired photons approaching the stations of Alice and Bob, with polarisers set at a generic relative angle configuration $(\mathbf{a}^*, \mathbf{b}^*)$. These constitute joint, marginal, and conditional probabilities, expressed as a formalism relevant to whatever the directional setup this may be. Let's look.

Joint probabilities:

$$P[(A(\mathbf{a}^*) = +1)(B(\mathbf{b}^*) = +1)] = P[(A(\mathbf{a}^*) = -1)(B(\mathbf{b}^*) = -1)]$$
$$= \tfrac{1}{2} \, cos^2(\mathbf{a}^*, \mathbf{b}^*) \, ,$$

and (1)

$$P[(A(\mathbf{a}^*) = +1)(B(\mathbf{b}^*) = -1)] = P[(A(\mathbf{a}^*) = -1)(B(\mathbf{b}^*) = +1)]$$
$$= \tfrac{1}{2} \, sin^2(\mathbf{a}^*, \mathbf{b}^*) \, .$$

Marginal probabilities:

$$P(A(\mathbf{a}^*) = +1) \qquad (2)$$
$$= P[(A(\mathbf{a}^*) = +1)(B(\mathbf{b}^*) = +1)] + P[(A(\mathbf{a}^*) = +1)(B(\mathbf{b}^*) = -1)]$$
$$= \tfrac{1}{2} \, cos^2(\mathbf{a}^*, \mathbf{b}^*) + \tfrac{1}{2} \, sin^2(\mathbf{a}^*, \mathbf{b}^*) = 1/2 = P(B(\mathbf{b}^*) = +1).$$

Conditional probabilities:

$$P[(A(\mathbf{a}^*) = +1)|(B(\mathbf{b}^*) = +1)] = cos^2(\mathbf{a}^*, \mathbf{b}^*)$$
$$\neq P[(A(\mathbf{a}^*) = +1)] = 1/2 \, ,$$

and (3)

$$P[(A(\mathbf{a}^*) = +1)|(B(\mathbf{b}^*) = -1)] = sin^2(\mathbf{a}^*, \mathbf{b}^*) \, , \text{ which is different.}$$

For efficiency in what follows, we shall denote the four probabilities appearing in equations (1) by $P_{++}, P_{--}, P_{+-},$ and P_{-+} when the pertinent angle setting is evident.

The four joint probabilities surely sum to equal 1, because the sum of $cos^2 + sin^2$ of any angle equals 1. A few properties of the probability mass function (pmf) they compose should be noticed. Firstly, the four probabilities can be specified by the value of any one of them. The equations (1) stipulate that no matter what the relative angle $(\mathbf{a}^*, \mathbf{b}^*)$ may be, the values of $P_{++} = P_{--}$, and $P_{-+} = P_{+-}$. Since the four probabilities do sum to 1, the specification of P_{++} as the value p, for example, implies that the pmf vector $[P_{++}, P_{--}, P_{+-}, P_{-+}]$ would be $[p, p, (1-2p)/2, (1-2p)/2]$. Moreover, an important feature of these equalities, which we shall discuss in detail shortly, is that the polarisation probabilities are symmetric, or "exchangeable". This is codified be the equality of the probabilities P_{+-} and P_{-+}.

Another feature of this quantum distribution is that the probabilities for the paired detection outcomes depend only on the *product* of the two measurements. For both outcomes $++$ and $--$ yield a product of $+1$, and both outcomes $+-$ and $-+$ yield a product of -1. Thus, the QM-motivated distribution for the experimental value of the polarisation *product* $A(\mathbf{a}^*)B(\mathbf{b}^*)$ is specified by

$$P[A(\mathbf{a}^*)B(\mathbf{b}^*) = +1] = cos^2(\mathbf{a}^*, \mathbf{b}^*)$$

and
$$P[A(\mathbf{a}^*)B(\mathbf{b}^*) = -1] = sin^2(\mathbf{a}^*, \mathbf{b}^*).$$

As will be important to recognise in what follows, the expected value of this detected product is

$$E[A(\mathbf{a}^*)B(\mathbf{b}^*)] = (+1)cos^2(\mathbf{a}^*, \mathbf{b}^*) + (-1)sin^2(\mathbf{a}^*, \mathbf{b}^*)$$
$$= cos\,2(\mathbf{a}^*, \mathbf{b}^*) \tag{4}$$

according to standard double angle formulas.

It is worthwhile reminding ourselves here that "the expected value of an observed quantity" is the "first moment" of its distribution. Geometrically, it is the point of balance of the probability mass function weights when they are positioned at the points on a number line where the observation might occur. It is a property of a probability distribution for the outcome of a specific single observable variable. It is not an average. Just saying.

The expectation value $E[A(\mathbf{a}^*)B(\mathbf{b}^*)]$ can also be represented as $\quad E[A(\mathbf{a}^*)B(\mathbf{b}^*)] = 2\,cos^2(\mathbf{a}^*, \mathbf{b}^*) - 1 = 4\,P_{++}(\mathbf{a}^*, \mathbf{b}^*) - 1, \quad$ (5) which will prove useful. For the value of $sin^2(\mathbf{a}^*, \mathbf{b}^*)$ appearing in the first equality of (4) can also be written as $1 - cos^2(\mathbf{a}^*, \mathbf{b}^*)$. Thus, the entire quantum distribution for the four possible polarisation observation pairs is also representable by the expectation of the polarisation product. Enough of this for now.

To conclude this discussion of the full joint distribution, it is now worth dwelling on the fact that the marginal probability that any photon passes through its polariser equals $1/2$, no matter the direction of the polariser it encounters. Generally, the joint probability that the two photons pass through their polarisers directed at $\mathbf{a}^*$ and $\mathbf{b}^*$ does not equal the product of the marginal probabilities for their passages. This result codifies a touted feature of physical processes at quantum scales of magnitude, that the photon behaviours of particle pairs are understood to be *entangled*. The conditional distribution for either one of the polarisation events depends on the context of the conditioning behaviour of the other:

$$P[(A(\mathbf{a}^*) = +1)|(B(\mathbf{b}^*) = +1)] \;=\; cos^2(\mathbf{a}^*, \mathbf{b}^*)$$
$$\neq\; P[(A(\mathbf{a}^*) = +1) \;=\; 1/2$$

and

$$P[(A(\mathbf{a}^*) = +1)|(B(\mathbf{b}^*) = -1)] \;=\; sin^2(\mathbf{a}^*, \mathbf{b}^*) ,$$

which is different still.

The standard interpretation of quantum physicists to these properties is that they certify the particles' behaviour as entangled. Thinking that a conditional probability $P(A|B)$ represents a change in the probability of event A given that the event B has been observed, they provocatively promote a contorted claim: that the observation of the particle at station B somehow changes the objective probabilistic behaviour of the particle at A, according to the touted probabilities discovered by quantum theory. For they think that quantum probabilities themselves govern the properties of matter at small scales, be it composed of photons or electrons or whatever. Moreover, they think that this feature of probabilistic structure is peculiar to quantum phenomena. Well, I assure you that none of this is true. It will only take a few pages to explain why ... decisively. I hope you are intrigued.

Before beginning the discussion of what *is* true, it is worth noticing one peculiarity of the Bell-defying argument. It is central to the propagation of this understanding of quantum entanglement, and crucial to the closure of all possible loopholes to the mistaken empirical defiance of Bell's inequality: specific polariser directions $\mathbf{a}^*$ and $\mathbf{b}^*$ are selected only *after* the entangled photons are ejected from their source, during the time interval before they meet their polarisers. Fair enough, as anyone can recognise the quantum probabilities

$$P[(A(\mathbf{a}^*) = +1)(B(\mathbf{b}^*) = +1)] \neq P(A(\mathbf{a}^*) = +1)\,P(B(\mathbf{b}^*) = +1)$$

for generic choice of the polariser directions at stations A and B. This identification of stochastic quantum dependence of the photon behaviours is said to certify their entanglement. ... But wait! Suppose that after the "entangled" photons leave their source, the directions $\mathbf{a}^*$ and $\mathbf{b}^*$ are chosen so as to constitute a relative angle $(\mathbf{a}, \mathbf{b}) = \pi/4$. With such a choice, quantum theory specifies

$$P[(A(\mathbf{a}) = +1)(B(\mathbf{b}) = +1)] = \tfrac{1}{2}cos^2(\pi/4) = 1/4\,.$$

But this is precisely the value the product of the marginal probabilities, $P(A(\mathbf{a}^*) = +1)\,P(B(\mathbf{b}^*) = +1)$, no matter what the angle $(\mathbf{a}^*, \mathbf{b}^*)$. So does this mean the photons have become *disentangled*, merely by our setting their polariser directions they will engage at the relative angle $\pi/4$? Yow! So they are entangled when they are ejected from their source, but they become disentangled during their journey, merely by our picking this specific relative polarisation angle?? I don't think so! Perhaps sad to say it, we will have to think some more.

You will recognise that I have already done a lot of computing in the production of this text. But no, I will not merely "shut up and compute", as has been provocatively advised. We will need to think aloud about two things that the popular physicists would rather avoid. These are the direct formal irrelevance of conditional probabilities to the *observation* of the conditioning quantity, and the symmetric feature of uncertain judgements, which do pertain to many, many matters of scientific interest, at all scales of magnitude. In a word, the assertion of conditional probabilities has nothing to do formally with the physical observation of the conditioning event. Moreover, symmetry is a feature of judgements that is not unique to quantum behaviour. These comments should colour your acquiescence to the proposal of quantum entanglement.

Perhaps to annoy the physicist further, we will need to think directly in terms of probability as the representation of scientific assertions with a modicum of uncertain knowledge, rather than as a property of natural matter. This was a mode of understanding natural to Einstein and to Schrödinger, and analysed to the hilt by Bruno de Finetti and associates. Let's go.

What does it mean, conditional probability?

Understood practically as the assertions of someone who is avowing the content of uncertain knowledge about the outcome of an experiment, all three of these types of probabilities we have enumerated (joint, marginal, and conditional) may be asserted simultaneously. In fact, coherency requires that the assertion of the joint probabilities be accompanied by the assertion of the marginal and both conditional probabilities. However, one can meaningfully assert only one or even both of the conditional probabilities, for example, without asserting the joint probability. The conditional probability assertion(s) would merely put conditions on what one might assert further as the joint or the marginal probabilities while preserving coherence.

Importantly, detector Bob does not have to observe the value of $B(\mathbf{b}^*)$ to make either of the two conditional probabilities specified in equations (3) to be operative. Neither of these depends on the observation of $B(\mathbf{b}^*)$ to have been engaged at all. *Both* of them can be asserted concomitantly, with substantive meaning. Notably, the theorem of total probability would ensure that, whatever the numerical values of the component assertions involved might be,

$$P(A(\mathbf{a}^*) = +1) = P[(A(\mathbf{a}^*) = +1)|(B(\mathbf{b}^*) = +1)] \ P(B(\mathbf{b}^*) = +1)$$
$$+ P[(A(\mathbf{a}^*) = +1)|(B(\mathbf{b}^*) = -1)] \ P(B(\mathbf{b}^*) = -1).$$

This statement relies on the joint assertion of *both the conditional probabilities*, conditioned on $B(\mathbf{b}^*) = +1$ and on $B(\mathbf{b}^*) = -1$, and both marginal probabilities as well.

The Copenhagen proposition that the quantum randomness of the polarisation behaviour of Alice's photon changes (the distribution collapses) when Bob observes his photon's behaviour as $B(\mathbf{b}) = +1$, for example, and changes differently (collapses onto a different projection) if he observes it as $B(\mathbf{b}) = -1$, ... and changes quite differently still if he observes not the value of $B(\mathbf{b})$ but rather $B(\mathbf{b}')$, ... misconstrues the role that probability plays in the formalisation of quantum mechanics.

Jocular and taunting discussions of whether the moon exists if we are not looking at it are a parody on a perfectly normal and common sense recognition. The experimental outcome at the paired polarisation setting $(\mathbf{a}', \mathbf{b}')$ just does not occur if we perform the experiment at the angle pairing $(\mathbf{a}, \mathbf{b})$ instead. Moreover, the photon approaching station A will engage its polariser whether or not anyone observes the polarisation phenomenon occurring at B. This is quite a different matter from deciding whether to look at the moon or not. The three types of probabilities characterised by equations $(1, 2, 3)$ are all meaningful and operationally defined assertions at any time. If a paired polarisation experiment is conducted on photons at the relative angle $(\mathbf{a}', \mathbf{b}')$, then the experiment at the relative angle $(\mathbf{a}, \mathbf{b})$ is just not happening. On the contrary, the moon's happening did involve beginning to shine at some point in time, continuing to shine now (no matter who is looking at it or not), and will stop shining in the very least when the sun stops shining, no matter what else is happening by then. Just by the way, in the meantime we'll be enjoying the spring tides and the neap tides, no matter who is looking. Smile.

The casual and suggestive language of Bayesian statistical reporting is somewhat to blame in promoting the misconception that conditional probability relies on the observation of the conditioning event. In this language, probabilities of the form $P(A(\mathbf{a}^*) = +1)$ and $P(B(\mathbf{b}^*) = +1)$ are commonly referred to as a "prior" probabilities, whereas one of the form $P[(A(\mathbf{a}^*) = +1)|(B(\mathbf{b}^*) = +1)]$ is commonly termed a "posterior" probability, ... posterior, that is, to the observation of $B(\mathbf{b}^*) = +1$. Such terminology may be harmless enough, and even evocative in the normal conduct of empirical analysis, when inference from designed experimental data or from scientific observation of natural history proceeds according to plan. However, it does not really tell the whole story of what is going on. The situation has been discussed extensively in foundational literature. It will take a little while to consider this, but let's try to digest it.

To be explicitly precise, notation for a prevision (probability) assertion, $P(E)$, ought to display a subscript designating the name of the person making the assertion, and the time at which the assertion is made. Even these would only be summary clues of the status of the assertion. One might also want to append the subscript with a formulation of what are the various components of the knowledge base of this person doing the asserting. Once we get to this level of detail, we could conceivably get lost in an unending description of the contextual basis of the assertion $P(\cdot)$. And we hardly want to. So we typically just write $P(\cdot)$, and get on with it.

Get on with what? Well, in fact we go through time formulating scientific assessments in terms of a sequence of such assertions regarding a vast array of vectors of observable quantities, sequential assertions we might formally denote by $P_{t-1}(\cdot), P_t(\cdot), P_{t+1}(\cdot)$ and so on where the "dot" in $P_t(\cdot)$ represents a vector of quantities and conditional quantities entertained at the time. Yes, scientific assertions do change with time, as we surely do well know. Not regularly from second to second, but surely from century to century, and from decade to decade, and at today's pace, even from day to day. Moreover, the assertions we make in one decade are quite typically not equal to the appropriate conditional probabilities we might have asserted ten years previously, simply conditioned on more observed data. We do continue to think!

So suppose we consider expressly such a sequence of uncertain assertions, and we write $P_t(A_t, B_t)$ to denote the shared array of uncertainty assertions of both Alice and Bob regarding the outcome of a polarisation experiment they are currently running. (We who are awaiting their observational reports may even share the same assertions too. Indeed, all who share the motivations of quantum theory share them too, Einstein included.) Now suppose that by the time the period $t+1$ eventuates, Bob has actually observed the value of $B_t(\mathbf{b'}) = -1$, for his photon say, but has not yet heard the result of the polarisation observation at Alice's. To simplify the following discussion, let's presume he has also been notified that Alice's polariser direction was settled at $\mathbf{a'}$ rather than $\mathbf{a}$, but he has not been informed of the value of $A_t(\mathbf{a'})$. The question now is, what is Bob's probability assertion $P_{B(t+1)}[A_t(\mathbf{a'}) = 1]$?

For an answer, we could rely on Bob's conditional prevision assertion, shared with Alice in the previous assessment period, that $P_t[A_t(\mathbf{a'}) = +1 | B_t(\mathbf{b'}) = -1] = sin^2(\mathbf{a'}, \mathbf{b'})$. Moreover, in times of "normal science", we would surely expect this to be his assertion at time $(t + 1)$ in this scenario. He would have imagined being in his currently $(t + 1)$-situational predicament of having observed $B_t(\mathbf{b'}) = -1$ as a possibility when he made the conditional assertion $P_t[A_t(\mathbf{a'}) = +1 | B_t(\mathbf{b'}) = -1]$. He may have asserted other conditional probabilities at time "t" as well, such as $P_t[A_t(\mathbf{a'}) = +1 | B_t(\mathbf{b'}) = +1]$, and/or $P_t[A_t(\mathbf{a'}) = +1 | B_t(\mathbf{b}) = -1]$. Operationally, these latter two amount to price valuations for tentative transactions that he now would know to have been *disengaged* on account of the observation we presume he has made,

$B_t(\mathbf{b}') = -1$. Well, Bob is now permitted to assert quite righteously

$$P_{\mathcal{B}(t+1)}[A_t(\mathbf{a}') = +1] = P_{\mathcal{B}(t)}[A_t(\mathbf{a}') = +1|B_t(\mathbf{b}') = -1] = sin^2(\mathbf{a}', \mathbf{b}').$$

Everyone would be pleased. He would have inferred from his observation exactly what we would have expected him to. But do notice that there is nothing about the axiomatic logic of subjective probability (or of formalist probability either) that *requires* this. To make a new assertion in time in this way has been named an act of *confirmational commitment*, in the works of Isaac Levi (1980) ... to commit yourself to an actual inference of what you had asserted during a previous period of uncertainty formalisation, as one conditional inference among many. See also Levi (1978) and Kyburg (1980). This is an activity that has been studied technically by Goldstein (1983, 1985) in his assessment of a coherency condition. While $P_{t+1}(X_{t+1})$ may differ from $P_t(X_{t+1}|X_t = x_t)$ after the value of X_t has been observed to equal x_t, it is coherency with the attitude of normal science that requires us to assert a prevision at time t that $P_t[P_{t+1}(X_{t+1})] = P_t(X_{t+1})$. This is a realistic form (not merely tautological) of asserting that your expectation for a quantity must equal your expectation for your future expectation of the quantity.

Bruno de Finetti is sometimes belittled for his refined attention to philosophical details regarding the meaning of mathematical constructions. But in so many contexts, the distinctions upon which he insisted have substantive merit. This was one of them. A conditional probability is *not* a re-evaluation of a probability, but rather a meaningfully distinct assertion of its own.

The bottom line of this interlude is that conditional probabilities do not really have anything to do routinely with the *observation* of the conditioning event ... or with the supposed collapse of a prospective propensity.

The next aspect of probability assertions that needs discussion is a common feature of scientific judgements regarding many phenomena, the characteristic of symmetry, or "exchangeability". In the quantum realm it has been objectified and codified as the entanglement of elementary particles. Let's think.

Symmetry, exchangeability, entanglement

Standard discussions of quantum theory make regular allusion to the stability of measured values in repetitions of experiments, a feature said to be crucial to the conduct of science. Jaeger (2009, p 79) calls this "the repeatability hypothesis". It is said that if we make a spin measurement, say of $A(\mathbf{a}^*)B(\mathbf{b}^*) = -1$, for which the measurement does not disturb the system, then if this measurement were to be repeated a short time later, it would necessarily be the same.

In the tradition stemming from Heraclitus, such musing is silly. All is flux, as the saying goes. Within the subjectivistic construct of probability, the sensible way to think about what underlies the vacuous claim to repeatability of observations is the technical feature of symmetry, which is commonly relevant to uncertain judgements. We are living in the midst of an explosion. The very notion of the repetition of an experiment under identical conditions is fraudulent. There are no repetitions, and there are no identical conditions. From early on in his evaluation of probability theory, de Finetti balked at the common verbiage of statistical reasoning ... "Consider a sequence of N trials of the same experiment conducted under identical conditions ..." Each observed event, each recorded quantity is its own unique happening, and should be treated as such in the formalisation of our uncertainty about it. The feature of judgements that codify symmetric scientific attitudes toward sequences of distinct experiments or other numerically encoded observations is termed "exchangeability". The term was coined following a suggestion of George Polya in discussion after a presentation "on the condition of partial equivalence" by de Finetti (1938) at a famous Conference on Probability in Geneva in 1937. Intriguing and challenging notes on all discussions there, along with a critical commentary, were prepared by de Finetti (1939).

Let's start with two different examples. Firstly, suppose the events E_1 and E_2 represent the identifications of whether I will eat cauliflower next Tuesday, and whether my petrol container in the shed has more than one litre of fuel in it right now. As it is, I don't know the answers to these questions, but I can affirm that my probability for the answers (yes, no) is larger than my probability for the answers (no, yes), and that my probabilities for both of these answer pairings exceed zero.

Alternatively, suppose that the events E_1 and E_2 identify whether each of two blooming dairy cows, Daisy and Molly, who have been inseminated with straws of sperm drawn from the same bull last week, will test positive for conception of a new life when I check them next week. Again, I do not know the answer to these questions, but I can aver that my probability valuation for (yes, no) is equal to my assertion for (no, yes). Of course Molly is a different animal than Daisy, each having distinctly different histories, and of course the sperm they have been served were different. Nonetheless, the situations in which they find themselves *appear to me* as notably similar in every dimension relevant to their fecundity that I can think of. My assertion of identical probabilities for the events $(\mathbf{E}_2 = (1, 0))$ and $(\mathbf{E}_2 = (0, 1))$ is a mark of the symmetry in my attitude toward them. About the petrol container and my eating cauliflower next week I profess no such symmetry. We are ready for some technicalities.

You are said to regard the events composing the vector $\mathbf{E}_N$ *exchangeably* if you assert equality of probabilities $P(\mathbf{E}_N = \mathbf{e}_N)$ and $P(\mathbf{E}_N = \mathbf{f}_N)$ whenever the two prospective observation sequences $\mathbf{e}_N$ and $\mathbf{f}_N$ are permutations of one another. If only one component of $\mathbf{E}_N$ were to equal 1 (the rest equalling 0), you would judge it equi-likely to be any one of them. If only two components were to equal 1, you judge it equi-likely that they would be any two, and so on. The same judgement would characterise your attitude toward any two strings of events with the same number of components equal to 1. We say that your assessments are *permutation symmetric*. Within the context of all judgements that regard such a vector of events exchangeably, there are many possibilities for coherent probability assessment of the number of sequence components that will equal 1 (and the remaining ones equal 0). In fact, an assessment of a probability mass function for the sum of the events will completely characterise an exchangeable distribution over the components of the event vector.

These technicalities are by now very well-known among probabilists, though unfortunately they are yet to filter down to introductory classes in probability and statistics. Even more disappointingly, the ideas are yet to filter down to the teachers of these classes, nor even to the majority of practising statisticians. If you are new to the idea, you may enjoy reading my pedagogical technical introduction in Lad (1996, Section 3.8, 170-188).

A note of emphasis is required. Notice that exchangeability is a feature of the uncertain judgement of the person who is asserting the probabilities. It is not the events (or quantities) in a sequence that *are* exchangeable, or might be *assumed to be* exchangeable. Molly and Daisy are two distinctly different animals, each with a life of her own. It is *you* who regards their possible conception results exchangeably, knowing perhaps not much more about them than that they are both heifers on a dairy farm, both being pure-bred Holstein stock served with sperm from the same bull. A concerned farmer, who might know of some difficulties having occurred during the early weeks of Daisy's birthing, might not regard the fecundity of these two heifers exchangeably. For someone to ask a question of whether a sequence of events is exchangeable or not is misplaced. The events are not exchangeable. They are distinct events. Some of them equal 1, some equal 0. You must answer the question for yourself, if you wish to, "do you regard these events exchangeably, or do you not?" Someone else in a different state of uncertain knowledge about them might answer the question differently than you do. Fair enough.

The point relevant to deliberations of quantum theoretic assessment of a sequence of paired spin experiments of any type, for example, is that the physically distinct activities at stations A and B will yield whatever specific pair of outcomes that they do. The four possibilities for the pair are $\{++, +-, -+, --\}$. Paradigmatic examples would be the results of Aspect's polarisation experiments on pairs of photons at various relative angle settings, or magnetic spin observations in Stern-Gerlach experiments on pairs of electrons. Rotational symmetries in the structure of various possible experimental designs assure that our probabilities for $+-$ and $-+$ observations depend only on the relative angle between the directions of the two polarisers or the pair of magnets. Our knowledge of the distances travelled by the two photons from their source to the polarisers is so crude relative to the precise wavelengths and rotational velocities (angular momentum) of the photons that we are clueless as to the precise angle at which either rotating photon meets its polariser. Our symmetric attitude toward their polarisation behaviours depends only the relative angle set between their polarisers, at the precision we know it.

Because we engage "to prepare the experimental setups in the same way" at the two detection stations as best we can, knowing nothing at all about any supplementary variables whose recognition might change this judgement, we regard them exchangeably. Such a judgement is commonly found useful for inferential statistical analysis in most every field of empiricism: physics, economics, biology, agriculture, medicine, sociology, or mental health. In almost every practical example of its applicability, the assessed exchangeability of two events entails the judgement that $P(E_2|E_1 = 1)$ differs from $P(E_2)$. There is nothing at all unusual about this.

I propose to you the following example of scientific observation at a classical scale that mimics the structure touted of quantum particles, considered to be mysteriously entangled. Let's join Dorothy back in Kansas. Smile. I should provide you with some detail that will help you to enjoy what we find there.

The state of Kansas in the USA is situated in the heart of the great plains of North America. Swept virtually flat by the receding glaciers of aeons past, and subsequently ploughed annually prior to planting, the uniformity of the topography, waving in wheat, is stunningly beautiful to a touring observer. The state alternates fairly regularly with North Dakota as the the state producing the largest volume of wheat. In both 2021 and 2022 the total acreage planted in winter wheat was some 7.3 million acres. (2.4 acres equals 1 hectare.) The land area of the state is 213,000 square kilometres. The average wheat yield per acre in 2021 was 52 bushels (a record harvest for the state) and in 2022 it was 38 bushels. The 14-year average yield to 2021 was 43 bushels per acre. One bushel of wheat weighs about 60 pounds (some 27.2 kilograms) or about one million kernels. Enough background detail.

Suppose we select two acres by lottery from the 7.3 million acres planted in wheat, labelling them as *Acre* **a** and *Acre* **b**. As the chips fall, the two acres are some 400 kilometres apart. Further, label their wheat yields in bushels in 2024 as Y_A and Y_B. Let

$$A(\mathbf{a}) = (Y_A \geq 43) - (Y_A < 43), \text{ and } B(\mathbf{b}) = (Y_B \geq 43) - (Y_B < 43).$$

(Parentheses around an inequality that is observed to be true denotes an event equal to 1, whereas if it is found to be false the parenthetic inequality equals 0.) Thus, both of these quantities may equal only $+1$ or -1. Knowing nothing more than this about these acres, an agronomist (and you or I) would regard the quantities exchangeably. Now suppose the agronomist asserts probabilities $P_{+-} = P_{-+} = .07$, and $P_{++} = P_{--} = .43$, saying "yes, wheat yields are fairly variable for the state from year to year, but they are not so variable from acre to acre within Kansas during a given year when standard industrial agricultural practices are followed." Several other knowledgeable agronomists are also uncertain about the values of $A(\mathbf{a})$ and $B(\mathbf{b})$, but they concur in such assertions. "Ho hum!" they say. "There is nothing unusual about this. We've seen it all before," they say, grinning. "Now if we knew something more about these particular two acres, such as their proximity to the several wildlife and bird refuges in the state, their history of fertiliser use, their specific detailed soil compositions, their positions relative to the nearest road, and the time series of their specific rainfall experience during the growing season, then I would be able to say something more precise about their yields in 2024."

"Holy smokes!" cries the physicist who overhears them. "Your probability assertions match precisely the entangled quantum probabilities for the polarisation behaviour of Alice and Bob's photons, with the relative polariser angle $(\mathbf{a}, \mathbf{b}) = -\pi/8$ radians."

The feature of exchangeability evidently governs both the quantum theoretic assessment of photon behaviour at the two detection stations and the agronomists' assessment of paired yields of wheat from selected acres of Kansas farmland. The probability assessments for permutations of the paired observation possibilities are identical in both cases. Nonetheless, the conditions at the two paired sites are different from one another in both cases. Otherwise they would be the same site! Despite their differences, quantum theory specifies an exchangeable assessment of their detection results. Whereas the probabilities for the path behaviours of the two photons depend on the relative angle between their polarising directions, there is nothing to favour a (yes, no) result over a (no, yes) result on account of the rotational symmetry of the situation with respect to the rotating angles of the photon waves. The two polarisation observations are regarded exchangeably. The same could be said of wheat yields at the two sites of Kansas. The only difference in the two paired situations is that agronomists are fairly sure of some further conditions relevant to the wheat growth, whereas physicists are still trying to delineate what further observable conditions may be relevant to the polarisation behaviour.

We should note that the shared probability distribution for both the quantum and classical scale paired observations is even more restricted than merely by exchangeability. Among the four possible observable outcomes $(-1, -1), (-1, +1), (+1, -1)$, and $(+1, +1)$, an exchangeable distribution over all four possible spin pair observations is characterised merely by the equality of the probabilities $P_{-+} = P_{+-}$. However, the form of our two example distributions over the quartet of possible outcomes is more restrictive still, because they specify that $P_{--} = P_{++}$ as well, specifying an even deeper form of symmetry to the assessment of the photons' polarisation behaviour. Though interesting, this feature will not detain us here. A final comment on exchangeability will conclude this discussion, allowing us to reconsider its characterisation as the entanglement of quantum particle behaviour.

The widespread understanding of the entanglement of the photon pair behaviour stems from the fact that the joint probability for any specific pair of behaviours is not equal to the product of their marginal probabilities. As the probabilities are typically understood to be ontic properties of the photons themselves in the context of the paired polariser directions, it is the "beings" of the photons that are thought to be entangled rather than independent of one another. Recognising the epistemic content of the probability assertions involved allows us to understand the entanglement rather differently, as a feature of our cognisance.

It is a standard feature of exchangeable distributions that they cannot generally be represented as the product of two marginal distributions. It is true that independent and identically distributed (iid) assessments of two events also specify an exchangeable distribution for the two of them together. However, the "iid" distributions constitute a very special subfamily of the family of exchangeable distributions. Furthermore, an assertion of independence is not usually appropriate to actual sequences of observable quantities, particularly when we want to collect statistics from some of them for the purposes of learning about others in the sequence. It is because we are eager to learn about the values of some of them based on what we find out about the others that we do not regard them independently.

It is our understanding of either photon's behaviour that is entangled with our understanding of that of the other's. Upon observation, the behaviour of the photon pair will be found to be represented by one of the four possibilities $(-1, -1), (-1, +1), (+1, -1)$, or $(+1, +1)$. Whichever of these exclusive results is observed, there is nothing entangled in the photons themselves. It is the probabilities (our uncertain assessments of the situation) that are entangled. The structure of the situation regarding the photons is no different from that motivating our exchangeable assessments of the fecundity of pairs of heifers who are surely not entangled, though they may be entangled with a bull. The peculiar histories of the two who engage a conception possibility are specific to each of them. It is we who regard the activities exchangeably, an attitude that can bear on phenomena both at the quantum scale and the classical scale.

It is time to deconstruct another notable feature of quantum probabilities, shared with probabilities for mundane phenomena.

Squares of complex amplitudes

It is commonly regarded in quantum theoretic circles that the mathematical derivation of complex-valued amplitude functions associated with the eigenvalues of measurement operators is a distinguishing feature of quantum behavioural analysis. Deliberations of quantum theory do not directly specify probability distributions over the possible results of a measurement. Rather, its considerations of rotationally symmetric activity generate the specification of a complex-valued amplitude function. Its square with its complex conjugate then yields the real-valued probability mass function relevant to the possibilities of experimental observations. Surprisingly perhaps, it can be made evident that there is nothing unique or exclusive to statistical quantum analysis about this situation either.

A provocative paper by F.H. Fröhner (1998), published in a relatively obscure German format by *Verlag Zeitschrift für Natureforschung*, has touted a theorem of Riesz-Fejer from the era of early quantum deliberations. It sounds startling, until you think about it. Riesz-Fejer showed that such a structure relating a complex wave function to a probability distribution function is inherent in *every* probability distribution! Rather than write it all up again myself, let me first merely quote from the abstract of the article "Missing Link Between Probability Theory and Quantum Mechanics: the Riesz-Fejer Theorem". Then I will display a brief example of the situation as relevant to a generic probability mass function over a finite domain. Finally, we shall examine the relevance of the theorem to the Gaussian distribution, so commonly used in applied statistics. Here is what Fröhner wrote in the abstract:

Quantum mechanics is spectacularly successful on the technical level but the meaning of its rules remains shrouded in mystery even more than seventy years after its inception. Quantum-mechanical probabilities are often considered as fundamentally different from classical probabilities, in disregard of the work of Cox (1946) – and of Schrödinger (1947) – on the foundations of probability theory. One central question concerns the superposition principle, i.e., the need to work with interfering wave functions, the absolute squares of which are probabilities. ... [In my present article] The superposition principle is found to be a consequence of an apparently little-known mathematical theorem for non-negative Fourier polynomials published by Fejer in 1915 that implies wave-mechanical interference already for classical probabilities. Combined with the classical Hamiltonian equations for free and accelerated motion, gauge invariance and particle indistinguishability, it yields all basic quantum features – wave-particle duality, operator calculus, uncertainty relations, Schrödinger equation, CPT invariance and even the spin-statistics relationship – which demystifies quantum mechanics to quite some extent.

You may read of the Fourier analysis in Fröhner's article for yourself if you wish. Here we shall merely examine the paradigmatic status of a common probability mass function, $\mathbf{p}_N = (p_1, p_2, ..., p_N)$, assessed for an observation limited to a finite and discrete ensemble of possibilities in the simplest cases of $\mathbf{p}_2$ and $\mathbf{p}_3$. After recognising these to equal the squares of associated complex wave functions, we shall conclude this Section with a simple display of the amplitude function for a Gaussian density function.

A simple probability, and a simple binomial mass function

To begin, it is evident that a wave representation of the components of the probability mass function for *any* event, $\mathbf{p}_2 = (p, 1 - p)$, can be specified by any pair of amplitudes $a_1 + b_1 i$ and $a_2 + b_2 i$, with their complex squares satisfying

$$p_1 = r_1^2 = (a_1 + b_1 i)(a_1 - b_1 i) = a_1^2 + b_1^2 ,$$

and
$$p_2 = r_2^2 = (a_2 + b_2 i)(a_2 - b_2 i) = a_2^2 + b_2^2 ,$$

along with the obvious restriction that $p_1 + p_2 = r_1^2 + r_2^2 = 1$. This representation of the pmf as a squared complex wave function is not unique, but is arbitrary relative to a phase factor, which we shall formalise below. This is a common situation recognised in quantum physics. See the didactic text of Susskind (2014, pp 22 and 108), for example. We can now appreciate that such a representation would be common to *any* application of probability, even though this infrastructure is not commonly required for recognition in most applications.

Consider a simple example in two dimensions relevant to any application at all, the mass vector $\mathbf{p}_2 = (.64, .36)$. Its wave representation by the coefficient pairs (a_1, b_1) and (a_2, b_2) could be specified by *any* two points on the disks $a_1^2 + b_1^2 = .64$ and $a_2^2 + b_2^2 = .36$, with radii of .8 and .6 respectively, as displayed in Figure 1. The

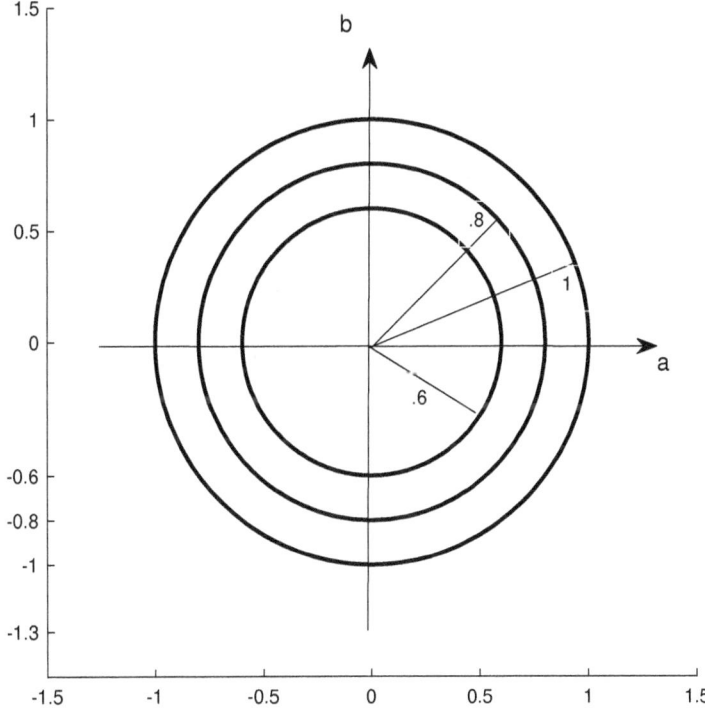

Figure 6: Any two pairs of complex amplitudes $a + bi$ for which $a_1^2 + b_1^2 = .64$ and $a_2^2 + b_2^2 = .36$ constitutes a wave representation of the probability mass vector $(.64, .36)$. Whereas the pairs $(a_1, b_1) = (.8, 0)$ and $(a_2, b_2) = (.6, 0)$ surely suffice, the complex rays $(r = .8, \ \theta = \pi/4)$ and $(r = .6, \ \theta = -7\pi/32)$ depicted here identify the pairs $(a_1, b_1) = (.5657, .5657)$ and $(a_2, b_2) = (.4638, -.3806)$. These would represent the wave function supporting this probability mass vector just as well.

simple choices of $a_1 = .8, b_1 = 0, a_2 = .6, b_2 = 0$ would identify the probabilities $(p_1, p_2) = (.64, .36)$ without any need to delve into the complex amplitude structure that underlies it, nor the many different choices of the complex (a, b) coefficients that it would permit. In most applications, such a choice is sufficient for all that is required. However, multiplying any two wave function components, each by a factor of $exp\{-i\gamma\} = cos(\gamma) + i\ sin(\gamma)$ using Euler's identity, say for an arbitrary positive value of γ, would merely rotate the wave function amplitudes around their circles of appropriate radius to another supporting wave representation of the probabilities. A wave component $(a + bi)$ would rotate to $[(a\ cos\gamma - b\ sin\gamma) + (b\ cos\gamma + a\ sin\gamma)\ i\ ]$, a complex vector with the same radius. The fact that quantum theoretic deliberations generate a probabilistic structure of measurement prognosis via Hermitian operators on vectors in a Hilbert space is what brings such structural underpinnings of the specified probability mass function to the fore of the theory.

An example one dimension higher will clinch the understanding. Consider the Binomial probability mass function $\mathcal{B}(n = 2, p = .64)$, with the three well-known components

$$\mathbf{p}_3 = (.36^2 = .1296,\ 2(.36)(.64) = .4608,\ .64^2 = .4096).$$

In standard complex representation of each wave amplitude by $a + bi$, any choices of three (a, b) vectors from discs with radii $.36, \sqrt{.4608}$, and $.64$ would suffice to represent the Binomial mass function $\mathbf{p}_3$ as the componentwise squares of the amplitude function. More generally, any array of points including one on each of three discs whose squared radii sum to 1 would identify a generic probability mass vector $\mathbf{p}_3$ in the unit-simplex in two dimensions. The binomial pmf for $n = 2$ constitutes a special case for which the relative sizes of the radii of the three disks are prescribed.

A final example pertinent to the Gaussian density function is worthy of display.

A standard Gaussian wave function

Consider a complex function of the form
$$g_+(x) \equiv (2\pi)^{-\frac{1}{4}} exp\{-\tfrac{1}{2}[(a + bi)x]^2\}$$
along with its conjugate function
$$g_-(x) \equiv (2\pi)^{-\frac{1}{4}} exp\{-\tfrac{1}{2}[(a - bi)x]^2\} .$$

The product of these two conjugate functions is

$$g_+(x)g_-(x) \;=\; (2\pi)^{-\frac{1}{2}}\,exp[-a^2x^2],$$

which for $a = \sqrt{\frac{1}{2}}$ yields the Gaussian density $\mathcal{N}(0,1)$ for any value of b. Other values of "a" would characterise a density with a different variance. Once again, the wave function representation of the Gaussian density is not unique, but has equivalent forms in terms of an arbitrary phase factor, the value of the complex coefficient b.

The article of Fröhner focuses on the curiously neglected history of the Riesz-Fejer result and the details of the Fourier representations involved. It does not provide a display of these simple examples, which I have generated from collaborative discussions with Diego Molteni at the University of Palermo. It is a simple matter to identify the amplitude functions associated with any parametric distribution. The insight they provide is that the wave form of a probability mass function and a probability density function are not at all peculiar to quantum probabilities, but rather are a structural characteristic of any probability distribution function at all.

Causation, properly relegated by Hume

Discussions of issues in quantum mechanics often revolve about conceptions of causal relations within physical processes, concerns which surely motivated Einstein in his consideration of local realism as a principle. In contrast, I consider it to be the conception of causality itself that is the source of confusion and misdirected research. Long relegated as the musings of an unusual outlier, the writings of David Hume are remarkably astute in asserting that the concept of cause is meaningless. Of course we use the word as an imaginary metaphor ... "if this had not happened then that would not have happened." However, formally, this means nothing as a claim about verifiable history. With respect to any two propositions, A and B, there are four confirmable conjunctive propositions: A and B, A and $\tilde{B}$, $\tilde{A}$ and B, and $\tilde{A}$ and $\tilde{B}$. There is no confirmable "A causes B". Be aware that the conclusion of every empirical study devised to study the "causes" of B ends with the statement, "Of course correlation is not causation, and we will need a lot more money for research to discover the causal links involved here." There is not enough money in the world. What we need to do is to reconceptualise our imagination of how to discover the glorious happening that is Natural History. What do we know has happened, and what do we think might happen next?

There are any number of further issues that might be of interest for discussion. However, my esteemed reader will be wondering now when it will all end. When is this guy going to shut up and start computing again? Well, that is what we shall do now. In the final chapter to conclude this book we shall engage a final computation that would have pleased John Bell.

Chapter 7

WHERE IT ALL STARTED ...
back to John Bell himself

To one who expects impassive rigour in the evaluation of scientific controversy, I find the folkways of the professional physics community unusual, to say the least, if not outlandish. Of course spirit and flair are to be welcomed. However as an outsider, it appears to me that the critical capacity of its institutions has succumbed to self-adulation of among its preeminent theoreticians. Apparent in Becker's (2018) documentation of the development of quantum theory in the twentieth century, the genius of Niels Bohr already was clouded by his inability to explicate his ideas clearly, even while inspiring devotion from followers who respected him highly and the Weltanschaung he ordained. More recently, in the long give and take of research publications regarding the Bell inequality, one finds a rather unusual exhibition of the syndrome. David Mermin (2005) proposed in a highly regarded article that an argument of Asher Peres need not have been defended in print when it had been challenged by Hess and Philipp (2004), merely because the esteemed Peres already "knew" the challenge was mistaken. Mermin presented a reply himself as a glowing tribute, making errors similar to those that have been resolved by the simulation study presented here in Chapter 5. John Bell would have had none of this, however. Though mistaken in his understanding of the defiance of his inequality, he patiently replied to challengers in his time. Moreover, he did not shrink from confronting even the eminent Albert Einstein and his colleagues in the article in which he first proposed the quantum defiance of his inequality.

With this background in mind, I shall dare conclude this book with a reassessment of the article of the duly honoured John Bell himself (1964) in which he first introduced his inequality in a commentary, "On the Einstein Podolsky Rosen paradox". Therein he proposed that insistence on local realism governing quantum phenomena would relegate its theoretical propositions as defiant of the inequality. The context in which he initially discovered this result involved a gedankenexperiment with a structure even simpler than the four-quantity CHSH format we assessed in Chapter 1, formalised just a few years later. It will become apparent once again that the conundrum Bell originally portrayed arose from the neglect of symmetric functional relations among the three quantities he considered.

Bell's conundrum regarding EPR

It is not unusual that the formalisation of intriguing scientific and mathematical results becomes simplified during the course of their discussion in the public forum of ideas. This has surely been the case during the past sixty years of discussion of Bell's inequality. The argument originally proposed by Bell (1964) used a different type of notation from that we have been following throughout this book. This has largely been Aspect's notation, as applied to the CHSH setup. In addition, Bell's challenge was formulated in the context of magnetic spin observations on a pair of electrons, rather than polarisation observations on a pair of photons. We have addressed this type of experimental setup already in Chapter 4, where the relevant quantum probabilities for paired spin observations appear on page 112. The most important specific that we shall use herein in this context is $E[A(\mathbf{a})B(\mathbf{b})] = -cos(\mathbf{a}, \mathbf{b})$, where $(\mathbf{a}, \mathbf{b})$ is the relative angle between the directions of the magnets at the stations A and B. To build on the understanding we have developed, I shall faithfully rephrase Bell's original result here using notation with which we are now familiar, and then assess it in these terms.

As originally proposed, Bell's inequality pertains to simultaneous spin-products for an electron pair, observed (in gedanken fashion) on a pair of electrons at only three relative magnet angle settings, $(\mathbf{a}, \mathbf{b}), (\mathbf{a}, \mathbf{c})$, and $(\mathbf{b}, \mathbf{c})$. Using notation in the style of Aspect we have devised in Chapter 1, Bell's inequality specifies

$$1 + E[A(\mathbf{b})B(\mathbf{c})] \geq \big| E[A(\mathbf{a})B(\mathbf{b})] - E[A(\mathbf{a})B(\mathbf{c})] \big|, \quad (15)$$

which Bell derived faultlessly, and numbered as his equation (15). The inequality involves expectations for three spin-products for a single pair of electrons. They are derivative from an expectation already evaluated over the values of posited "supplementary variables" λ, which Einstein had proposed as influential on activities at the observation stations A and B.

The inequality results jointly from quantum theoretic recognition of the empirical observation that magnets facing the same polar direction repel one other (implying $A(\mathbf{b}, \lambda) = -B(\mathbf{b}, \lambda)$), and the supposition of locality. This proposes that the activity recorded as $A(\mathbf{a}, \lambda)$ in any instance is not affected by the direction $\mathbf{b}$ or $\mathbf{c}$ of the magnet it is paired with at station B, nor with the spin exhibited by its paired electron there. Its value would be presumed to be identical in this instance when paired with either of them. This is despite the fact that the quantum probabilities for their joint occurrence are clearly recognised as conjoined.

Bell's first challenging claim then pertained to the particular setup in which the pertinent Stern-Gerlach magnet angles involve nearly identical magnet directions, $\mathbf{b}$ near to $\mathbf{c}$. His argument was casual, but apparently begins correctly. If the direction $\mathbf{b}$ is quite close to $\mathbf{c}$, then the right-hand side of his inequality (15) is near to the angular distance $|\mathbf{b} - \mathbf{c}|$, whereas it appears that the left-hand side is much smaller, *not larger* as the inequality requires.

As an example, suppose $(\mathbf{a}, \mathbf{b}) = .8888888$ (in units of radians) and $(\mathbf{a}, \mathbf{c}) = .8888887$, making $(\mathbf{b}, \mathbf{c}) = -.0000001$. Then it appears that the left-hand side of the inequality yields

$$LHS = 1 - cos(-.0000001) = 4.9960 \times 10^{-15},$$

while the right-hand side yields

$$RHS = |cos(.8888888) - cos(.8888887)| = 7.7637 \times 10^{-8},$$

which is a larger value to be sure. This RHS value is indeed quite close to $|\mathbf{b} - \mathbf{c}| = 1 \times 10^{-7}$.

If these *joint* evaluations were correct, the inequality would be defied. Bell merely remarks that $E[A(\mathbf{b})B(\mathbf{c})]$ cannot be stationary at its minimum value (which is -1 when $\mathbf{b} = \mathbf{c}$), and cannot equal the quantum mechanical value $-cos(\mathbf{b}, \mathbf{c})$. His allusion to non-stationarity pertains to the fact that if the magnet directions were actually identical ($\mathbf{b} = \mathbf{c}$) then the inequality (15) would merely say $0 \geq 0$, which *is* true; whereas if the paired magnet directions specify an angle merely close to 0 (however close) it remains false.

Well something must be wrong here, as Bell himself suspected in the quotation we highlighted in Chapter 1, but could not identify. What could it be? You should not be surprised by now to find that the problem was that there is an array of functional relationships among the spin-products $A(\mathbf{a})B(\mathbf{b})$, $A(\mathbf{a})B(\mathbf{c})$, and $A(\mathbf{b})B(\mathbf{c})$ in this evaluation of expectations for the three spin-products in a gedankenexperiment, which has been neglected. All three of these expectations appear in the two sides of the inequality. Recognising the functional relations involved in the production of the three gedanken spin-products will put paid to the defiance of the inequality. Let's look into it.

Bell's error of neglect

Without fanfare, let's merely examine the realm matrix of possible gedanken outcomes of the three spin-products for the pair of electrons under consideration. We are practised by now in such constructions.

$$
\mathbf{R}
\begin{pmatrix}
A(\mathbf{a}) \\
B(\mathbf{b}) \\
B(\mathbf{c}) \\
A(\mathbf{b}) \\
**** \\
A(\mathbf{a})B(\mathbf{b}) \\
A(\mathbf{a})B(\mathbf{c}) \\
A(\mathbf{b})B(\mathbf{c}) \\
**** \\
1 + A(\mathbf{b})B(\mathbf{c}) \\
A(\mathbf{a})B(\mathbf{b}) - A(\mathbf{a})B(\mathbf{c})
\end{pmatrix}
=
\begin{pmatrix}
1 & 1 & 1 & 1 & * & -1 & -1 & -1 & -1 \\
1 & 1 & -1 & -1 & * & 1 & 1 & -1 & -1 \\
1 & -1 & 1 & -1 & * & 1 & -1 & 1 & -1 \\
-1 & -1 & 1 & 1 & * & -1 & -1 & 1 & 1 \\
* & * & * & * & * & * & * & * & * \\
1 & 1 & -1 & -1 & * & -1 & -1 & 1 & 1 \\
1 & -1 & 1 & -1 & * & -1 & 1 & -1 & 1 \\
-1 & 1 & 1 & -1 & * & -1 & 1 & 1 & -1 \\
* & * & * & * & * & * & * & * & * \\
0 & 2 & 2 & 0 & * & 0 & 2 & 2 & 0 \\
0 & 2 & -2 & 0 & * & 0 & -2 & 2 & 0
\end{pmatrix}
$$

To begin its analysis, the first three rows of this realm matrix constitute the Cartesian product $\{-1, +1\}^3$, as is appropriate to logically independent quantities. Each of the three spin values can be either -1 or $+1$. However, the conditions of the experiment require that spin values of electrons facing magnets in the same direction at stations A and B be opposed. So spin values in the fourth row of the matrix equal the negative of those in the second row. The values of $A(\mathbf{b}) = -B(\mathbf{b})$ in every column of the top partition. This is required of repelling magnets facing the same polar direction, as was recognised by Bell when deriving the inequality.

Now we have noticed the first striking feature of this realm matrix before. While there are eight distinct columns provided by the first three spin possibilities in the top partition, there are only four distinct columns in the second and third row partitions involving spin-products, the final five rows of the matrix. That is why the matrix display is partitioned vertically. Only the left half of the matrix is relevant to an analysis of expectations involving the final two blocks, which would be involved in assessing Bell's original inequality according to the prognostications of quantum theory.

The second feature of the matrix, most importantly, pertains to the *spin-products* of the four component spins. Their concomitant possibilities are enumerated in the second vertical partition of the realm. The second row of the second horizontally partitioned block turns out to equal negative the component products of its first and third rows. Bell had noticed this in his algebraic derivation that

$$A(\mathbf{a})B(\mathbf{c}) = -A(\mathbf{a})B(\mathbf{b})A(\mathbf{b})B(\mathbf{c}) , \quad \text{since } B(\mathbf{b})A(\mathbf{b}) = -1.$$

However, he subsequently neglected to account for the full extent of symmetric functional relations this engenders among the spin-products. When he evaluated separately the expectations of the three products appearing in his inequality, he ignored some pertinent features. The result he did recognise identifies the numerical value of $A(\mathbf{a})B(\mathbf{c})$ as restricted to equal a function value of the products $A(\mathbf{a})B(\mathbf{b})$ and $A(\mathbf{b})B(\mathbf{c})$: it is negative the value of *their* product. In fact however, *every row of the second block partition of the realm matrix is restricted symmetrically to equal the same function value of the other two row, the negative product of their components!* Look at the three rows of the second vertical partition, and you will see this.

Algebraically, $A(\mathbf{a})B(\mathbf{b}) = -A(\mathbf{a})B(\mathbf{c})A(\mathbf{b})B(\mathbf{c})$,
$$\text{since } B(\mathbf{c})B(\mathbf{c}) = 1, \text{ and } A(\mathbf{b}) = -B(\mathbf{b}).$$

Furthermore,
$$A(\mathbf{b})B(\mathbf{c}) = -A(\mathbf{a})B(\mathbf{b})A(\mathbf{a})B(\mathbf{c}) ,$$
$$\text{since } A(\mathbf{a})A(\mathbf{a}) = 1, \text{ and } B(\mathbf{b}) = -A(\mathbf{b}).$$

These are the two relations whose recognition eluded Bell.

The upshot of these functional relations is that an assessment of quantum expectations for the final two rows of block three must take them all into account. Once again, quantum theory does not specify a complete joint distribution for the three gedanken spin-products. If quantum expectations were entertained for any two of them, then only bounds can be computed for the cohering expectation of the third, using linear programming methods of the type we have used regularly throughout this book. Choice of two expectations to constitute linear restrictions for the computation can be done in three ways.

To specify the required computational programs here, and to report the results, the three rows of the second partitioned block of the realm matrix will be designated with the names $\mathbf{r_{ab}}, \mathbf{r_{ac}}$, and $\mathbf{r_{bc}}$. These will be used in pairs when specifying linear programming constraints in computations for which the objective function is one of the final two rows of the realm matrix. These rows will be designated as $\mathbf{r_{LHS}}$, and $\mathbf{r_{RHS}}$, respectively. Specifying precise quantum expectations for each choice of two of the spin-products identifies one of the expectations appearing in block three, leaving the other to be constrained merely by the computed bounds. By computing these bounds, we can examine the implications of the results for the questionable adherence of quantum probabilities to Bell's inequality in the form he originally proposed it. The application of the absolute value evaluation of the right-hand side of the inequality will be recognised in the discussion.

Bounding results pertinent to Bell: quantum satisfaction of the inequality?

To begin, consider the relative magnet angles again on the order at which Bell thought his inequality was defied, when the difference between the magnet directions $\mathbf{b}$ and $\mathbf{c}$ is small: specifically, again at $(\mathbf{a}, \mathbf{b}) = .8888888$ (in units of radians) and $(\mathbf{a}, \mathbf{c}) = .8888887$. Then the quantum theoretic evaluation of the right-hand side of Bell's inequality, an absolute value, equals

$$\left| -cos(.8888888) + cos(.8888887) \right| = 7.763718345987769 \times 10^{-8}.$$

Concomitantly, only bounds can be computed for the quantum theoretic evaluation of the left-hand side of the inequality that agree with these assessments, according to the linear programs

$$\binom{min}{max} 1 + E[A(\mathbf{b})B(\mathbf{c})] = \mathbf{r_{LHS}}\, \mathbf{q_4} \ ,$$

subject to the linear constraints

$$\mathbf{r_{ab}}\, \mathbf{q_4} = -cos(.8888888) \quad \text{and} \quad \mathbf{r_{ac}}\, \mathbf{q_4} = -cos(.8888887),$$

along with the unit constraint $\mathbf{1_4^T}\mathbf{q_4} = 1$ on non-negative $\mathbf{q_4}$.

These yield the bounding interval for the objective function $r_{LHS} q_4$ that coheres with quantum probabilities as the large interval

$$(7.763718368192229 \times 10^{-8}, \quad 0.739449682495256).$$

Every number in this interval motivated by quantum theoretic probabilities exceeds the right-hand side of Bell's inequality, from the sixteenth decimal place! **We've got inequality satisfaction!**

But there is still more satisfaction to be garnered!

Suppose we presume quantum probabilities for spin-products at two alternate related magnet angles at which Bell thought his inequality was defied, $(\mathbf{a}, \mathbf{b}) = .8888888$ (in units of radians) and $(\mathbf{b}, \mathbf{c}) = -.0000001$. Then the left-hand side of Bell's inequality equals $1 - cos(-.0000001) = 4.996003610813204 \times 10^{-15}$, quite a small number, to be sure.

However, the bounds on the right-hand side of the inequality that is enclosed in the absolute value operator, as computed via the linear programming setup

$$\left(\begin{smallmatrix}min\\max\end{smallmatrix}\right) E[A(\mathbf{a})B(\mathbf{b})] - E[A(\mathbf{a})B(\mathbf{c})] = r_{RHS} q_4$$

subject to the linear constraints

$$r_{ab} q_4 = -cos(.8888888) \quad \text{and} \quad r_{bc} q_4 = -cos(-.0000001),$$

along with the unit constraint $1_4^T q_4 = 1$

are that it lie within ... the interval $[0,0]$. (!) The right-hand side of the inequality prescribed by quantum theoretic probabilities under these conditions *must equal* 0. This is to say, that when the directions of the magnet settings $\mathbf{b}$ and $\mathbf{c}$ are so near to one another, then the expectations $E[A(\mathbf{a})B(\mathbf{b})]$ and $E[A(\mathbf{a})B(\mathbf{c})]$ must be identical. Again the left-hand side surely exceeds the right-hand side of Bell's inequality, which must equal 0. **Again, we've got satisfaction of Bell's inequality!**

Finally, there is complete satisfaction! Consider the final two other relative magnet angles among these three, at which Bell thought his inequality was defied, $(\mathbf{a}, \mathbf{c}) = .8888887$ (in units of radians) and $(\mathbf{b}, \mathbf{c}) = -.0000001$. This is essentially the same problem we have just assessed. In this case the left-hand side of the inequality again equals

$$1 - cos(-.000001) = 4.996003610813204 \times 10^{-15}.$$

By now not surprisingly, the bounds on the right-hand side of the inequality, which is enclosed in an absolute value operator, are identical. These derive from the linear programming computations

$$\binom{min}{max} \quad E[A(\mathbf{a})B(\mathbf{b})] - E[A(\mathbf{a})B(\mathbf{c})] = \mathbf{r_{RHS}} \, \mathbf{q_4}$$

subject to the linear constraints

$$\mathbf{r_{ac}} \, \mathbf{q_4} = -cos(.8888887) \quad \text{and} \quad \mathbf{r_{bc}} \, \mathbf{q_4} = -cos(-.0000001),$$

along with the unit constraint $\mathbf{1_4^T} \mathbf{q_4} = 1$. The conditions of this problem are essentially the same. Once again, the objective function must lie within ... the interval $[0,0]$. The inequality is satisfied *whenever* the direction of the magnet pairing $(\mathbf{a},\mathbf{b})$ is very close to that of $(\mathbf{a},\mathbf{c})$.

Every coherent application of quantum probabilities to Bell's considered gedankensetup satisfies his inequality. Thus, the spectre of supposed instability of the quantum solution when the two angle pairings are identical, $(\mathbf{b} = \mathbf{c})$, does not arise. The applicable quantum expectations then yield the truism $0 \geq 0$, satisfying the inequality continuously as they do while $\mathbf{b}$ approaches $\mathbf{c}$.

Bell's inequality is satisfied with locality

Hardly defying Bell's inequality, the probabilistic prescriptions of quantum theory provide complete satisfaction of Bell's inequality.

Not a one to get his dander up when he hears of these results, John Bell would be pleased, I am sure.

REST IN PEACE, JOHN BELL !

... not to speak of EPR, whose viewpoint on the incompleteness of quantum theory had been undermined by Bell's mistaken proposition regarding quantum theoretic defiance of his inequality. The investigations of supplementary variables may continue without suspicion. You may continue to appreciate the moon and the tides, even while you are not actively observing them.   Smile.

References

Abbott, E.A. (1884) *Flatland: a romance of many dimensions*, London: Sealey and Co. Many recent editions.

Adenier, G. (2001) A refutation of Bell's theorem, in *Foundations of Probability and Physics*, Khrennikov, A. (ed.), pp. 29-38.

Aschwanden, M., Philipp, W., Hess, K., and Barraza-Lopez, S. (2006) *Quantum Theory, Reconsideration of Foundations*, 3, AIP Conference Proceedings, 437- 446.

Aspect, A. (2002) Bell's Theorem: the naive view of an experimentalist, in *Quantum [Un]speakables — from Bell to Quantum Information*, R.A. Bertlmann and A. Zeilinger (eds.), Springer.

Aspect, A., Grangier, P., and Gérard R (1981) Experimental tests of realistic local theories via Bell's theorem, *Physical Revue Letters* **47** (8): 460–3.

Aspect, A., Dalibard, J., and Gérard R (1982) Experimental tests of Bell's inequalities using time-Varying analyzers, *Physical Revue Letters* **49** (25): 1804-7.

Bassi, A. (2023) Orcid ID orcid.org/0000-0001-7500-387X, (2020)

Becker, Adam (2018) *What is real? The unfinished quest for the meaning of quantum physics*, New York: Basic Books, 444pp.

Bell, J.S. (1964) On the Einstein Podolsky Rosen paradox, *Physics*, 1(3), 195-200.

Bell, J.S. (1966) On the problem of hidden variables in quantum mechanics, *Reviews of Modern Physics*, **38**(3), 447-452.

Bell, J.S. (1971) Introduction to the hidden variables question, reprinted in Bell, J.S. (1987), 29-39.

Bell, J.S. (1975) Locality in quantum mechanics: reply to critics, reprinted in Bell, J.S. (1987), 63-66.

Bell, J.S. (1981) Bertlmann's socks and the nature of reality, reprinted in Bell, J.S. (1987), 139-158.

Bell, J.S. (1987) *Speakable and Unspeakable in Quantum Mechanics*, Cambridge Univ Press.

Bruno, G., and Giglio, A. (1980) Applicazione del metodo del simplesso al teorema fondamentale per la probabilità nella concezione soggettivistica, *Statistica* **40**(3), 337-344.

Brunner, N., Cavalcanti, D., Pironio, S. and Scarani, V. (2014) Bell nonlocality, *Reviews of Modern Physics*, **86**(2), 420-478.

Capotorti, A., Lad, F., and Sanfilippo, G. (2007) Reassessing accuracy rates of median decisions, *The American Statistician* **61**(2), 132-138.

Clauser, J.F., Horne, M.A., Shimony, A., and Holt, R.A. (1969) Proposed experiment to test local hidden-variable theories, *Physical Review Letters*, **23**, 880.

Cox, F. (1946) Probability, Frequency and Reasonable Expectation, *American Journal of Physics* **14**(1), 1-13.

Einstein, A., Podolsky, B, and Rosen, N. (1935) Can quantum mechanical description of physical reality be considered complete? *The Physical Review*, **47**, 777-780.

Falk, D. (2016) New Support for alternative quantum view, *Quanta magazine*, May issue.

Fine, A. (1982) Hidden variables, joint probabilities, and the Bell inequalities, *Physical Review Letters*, **48**(5), 291-294.

de Finetti, B. (1938) Sur la condition d'equivalence partielle, *Actualités Scientifique et Industrielle #* 739, Paris: Hermann. English translation in *Studies in Inductive Logic and Probability*, vol II, R. Jeffrey (ed.), P. Benacerraf and R. Jeffrey (trs.), Berkeley: University of California Press.

de Finetti, B. (1939) Compte rendu critique du Colloque de Genève sur la theorie des probabilités, *Actualités Scientifique et Industrielle #* 766, Paris: Hermann.

de Finetti, B. (1972) A useful notation, in *Probability, Induction and Statistics*, New York: John Wiley, pp. xviii-xxiv; translated by L. J. Savage from de Finetti (1967), Quelque conventions qui semblent utile, *Revue Roumaine de Mathématiques Pures et Appliqués*, **12**, 1227-1223.

de Finetti, B. (1972) *Probability, Induction and Statistics*, New York: John Wiley, which includes the translations "On the Axiomatisation of Probability", pp 67-114; (tr) G. Majone, from Sull'impostazione assiomatica del calcolo delle probilità, *Annali Triestini dell'Università di Trieste*, 1941, vol xix, pp 29-81 and "On the Abstract Theory of Measure and Intregration", pp 115-128. Sulla teoria astratta della misura e dell'integrazione, *Annali di Matematica Pura ed applicata* Series IV, Vol XL, pp 307-320, 1955

de Finetti, B. (1974, 1975) *Theory of Probability*, 2 volumes, A.F.M. Smith and A. Machi (trs.), New York: John Wiley.

Freire Junior, O. (2015) *Quantum Dissidents: rebuilding the foundations of quantum mechanics (1950-1990)*, Springer.

Fröhner, F.H. (1998) Missing Link Between Probability Theory and Quantum Mechanics: the Riesz-Fejer Theorem, *Zeitschrift Naturforschung*, **53 a**, 637-654.

Gigerenzer, G., Swijtink, Z., Porter, T., Daston, L., Beatty, J., and Krüger, L. (1989) *The Empire of Chance: how probability changed science and everyday life*, Cambridge: Cambridge University Press.

Gill, R. (2014) Statistics, Causality and Bell's Theorem, *Statistical Science* **29**(4), 512–528.

Gill, R. (2003) Time, finite statistics, and Bell's fifth position. In *Foundations of Probability and Physics*, **2**.

Goldstein, M. (1983) The prevision of a prevision, *JASA*, **78** 817-819.

Goldstein, M. (1985) Temporal coherence, in *Bayesian Statistics 2*, Bernardo, J.M. et al (eds), Amsterdam: North Holland, 231-248.

Greenberger, D. M., Horne, M. A., and Zeilinger, A. (1989) in *Bell's Theorem, Quantum Theory, and Conceptions of the Universe*, Kafatos, M. (ed), Dordrecht: Kluwer Academic, 73-76.

Greenberger, D.M., Horne, M.A., Shimony, A., and Zeilinger, A. (1990) Bell's Theorem without inequalities, *American Journal of Physics*, **58**(12), 1131-1143.

Greenstein, G., and Zajonc, A. (2006) *The Quantum Challenge: modern research on foundations of QM*, Second edition, Sudbury, Mass: Jones and Bartlett.

Hacking, I. (1975) *The Taming of Chance*, Cambridge: Cambridge University Press.

Hensen, B., Bernien, H, Dréau, A.E., and a list (2015) Loophole-free Bell inequality violations using electron spins separated by 1.3 kilometers, *Nature*, **526**,682-686.

Hess,K. and Philipp, W. (2004) Breakdown of Bell's theorem for certain objective local parameter spaces, *Proceedings of the National Academy of Sciences*, **101**, 1799–1805.

Hess K. (2014) Einstein Was Right! Singapore: Pan Stanford.

Jaeger, G. (2009) *Entanglement, Information, and the Interpretation of Quantum Mechanics*, Berlin, Heidelberg: Springer-Verlag.

Jaynes, E.T. (1986) Predictive Statistical Mechanics, in Moore G.T., Scully M.O. (eds) *Frontiers of Nonequilibrium Statistical Physics*, NATO ASI Series (Series B: Physics), **135**.

Kolmogorov, A.N. (1933) *The Foundations of Probability*, N. Morrison (tr.) New York: Chelsea.

Kyburg, H. (1980) Conditionalisation, *Journal of Philosophy*, **77**, 98-114.

Lad, F. (1996) *Operational Subjective Statistical Methods: a mathematical, philosophical and historical introduction*, New York: John Wiley.

Lad, F. (2020) The GHSZ argument: a gedankexperiment requiring more denken, *Entropy*, **22**(7), 759-ff.

Lad, F. (2021) Violation of Bell's inequality: a misunderstanding based on a mathematical error of neglect, *Journal of Modern Physics*, **12**, 1109-1144.

Lad, F. (2021b) Quantum mysteries for no one, *Journal of Modern Physics*, **12**, 1366-1399.

Lad, F. (2022) Resurrecting the prospect of supplementary variables with the principle of local realism *Applied Mathematics*, **2**, 159-169.

Lad F. (2023) Further investigations of the Aspect/Bell error: maximum entropy assessment, *Journal of Modern Physics*, **14**, 1272-1285.

Lad, F., Dickey, J.M., and Rahman, M. (1990) The fundamental theorem of prevision, *Statistica*, **50**, 19-38.

Levi, I. (1978) Confirmational conditionalisation, *Journal of Philosophy*, **75**, 730-737.

Levi, I. (1980) *The Enterprise of Knowledge: an essay on credal knowledge, probability, and chance*, Cambridge: M.I.T. Press.

Mermin, N. David (1991) Quantum mysteries for anyone, in *Boojums all the way through: communicating science in a prosaic age*, Cambridge: Cambridge University Press, and elsewhere.

Mermin, N.D. (1993) Hidden variables and two theorems of John Bell, *Reviews of Modern Physics*, **65**, 803-815.

Mermin, N.D. (2005) What's wrong with this criticism?, *Foundations of Physics*, **35**(12), 2073-2077

Mermin, N.D. and Schack, R. (2018) Homer nodded: von Neumann's surprising oversight, *Foundations of Physics*, **48**(9), 1007-1020.

von Mises, R. (1957) *Probability, Statistics, and Truth*, Hilda Geiringer (tr.), 2nd edition, London: Allen and Unwin.

von Mises, R. (1964) *Mathematical Theory of Probability and Statistics*, edited and complemented by Hilda Geiringer (tr.), New York: Academic Press.

von Neumann, J. (1932, English translation 1955) *Mathematical Foundations of Quantum Mechanics*, R. Beyer [tr], Princeton University Press.

Pan, J-W., Bouwmeester, D., Daniell, M., and Zeilinger, A. (2000) Experimental entanglement purification of arbitrary unknown states, *Nature*, **423**, 417-422.

Phillips, W.D. and Dalibard, J. (2023) Experimental tests of Bell's inequalities: a first-hand account by Alain Aspect, *The European Physical Journal D*, **77**:8, 1-14.

Redhead, M. (1987) *Incompleteness, nonlocality, and realism: a prolegomenon to the philosophy of quantum mechanics*, Oxford: Clarendon Press.

Reichenbach, H. (1949) *The theory of probability : an inquiry into the logical and mathematical foundations of the calculus of probability* , E.H. Hutten and Maria Reichenbach (trs.) 2nd edition, Berkeley: Univ California Press.

Reichenbach, H. (1944) *Philosophic Foundations of Quantum Mechanics*, Berkeley: Univ California Press.

Santilli, R.M. (1996) Nonlocal-Integral Isotopies of Differential Calculus, Mechanics and Geometries, *Rendiconti Circolo Matematico Palermo*, Suppl. Vol. 42, p. 7-82.

Savage, L.J. *Foundations of Statistics* (1954) New York: John Wiley.

Schrödinger, E. (1947) The Foundation of the Theory of Probability: I *Proceedings of the Royal Irish Academy*. Section A: Mathematical and Physical Sciences **51**, 51-66, and part II, 141-146.

Scozzafava, R. (2000) The role of probability in statistical physics, *Transport Theory and Statistical Physics*, **29**, 107 123.

Whittle, P. (1970) *Probability*, Harmondsworth, Middlesex: Penguin.

Whittle, P. (1971) *Optimization under Constraints*, London: Wiley.

Wiseman, H. (2015) Death by experiment for local realism, *Nature*, **526**, 649-650.

Zabell, S. (2009) De Finetti, chance, quantum physics, in *Bruno de Finetti, Radical Probabilist*, Maria Gavalotti (ed.), Berlin: Springer-Verlag.

www.ingramcontent.com/pod-product-compliance
Lightning Source LLC
Chambersburg PA
CBHW060833170526
45158CB00001B/155